Physical Methods in Chemistry and Nano Science.
Volume 3: Chromatography

Physical Methods in Chemistry and Nano Science.
Volume 3: Chromatography

Editor

Andrew R. Barron

Contributors

Andrew R. Barron, Alejandra Garcia Piantanida, Sha Li,
Matthew Makansi, Mahnoor Rafik Madakia,
Mustafa Salih Hizir, Pavan M. V. Raja,
Brett Virgin-Downey, Zhun Zhao

MiDAS Green Innovations
2020

MiDAS Green Innovation, Ltd
Swansea, SA1 8RD, UK

www.midasgreeninnovation.com

Dedication

To all my postgraduate researchers, I hope I gave your careers the start they deserved.

"To a brand new day, like the whole world has been made again"
Marillion (1994)

Contents

Acknowledgements

I would like to thank all the contributors to this Volume, for their interest in creating a user-friendly text for their peers.

The myriad collaborators of whom I have had the pleasure of learning from over the years are to be thanked for bringing new characterization methods to my research - you know who you are.

Last, but not least, I would like to thank my wife, Merrie, for putting up with me during the Editing of this book while 'social distancing' during the COVID-19 pandemic.

Preface

This Series intended as a survey of research techniques used in modern chemistry, materials science, and nanoscience. The topics are grouped into volumes, not be method *per se*, but with regard to the type of information that can be obtained. Thus, the Volumes are ordered as follows:

- Elemental composition.
- Physical and thermal analysis.
- Chromatography
- Chemical speciation.
- Molecular and solid state structure.
- Surface morphology and structure at the nanoscale.
- Device performance.
- Applications of analytical methods

As a consequence of this organization methods can be found in different Volumes. For example, X-ray photoelectron spectroscopy is included under Elemental Composition (Volume 1) with regard to its use for determining the chemical composition, while it is included under Chemical Speciation (Volume 3) with regard to determining the identity of component chemical moieties.

The goal was to create simple to understand explanations of methods that allow the reader to gain the knowledge to correctly apply a technique or interpret data. As a consequence, the topics in this book have been developed in partnership with undergraduate and postgraduate students at Rice University over a 7-year period, and because of this there is some variation in depth and focus given to each topic. I make no apology for this diversity.

Chapter 1: Gas Chromatography

Alejandra Garcia Piantanida, Zhun Zhao and Andrew R. Barron

Introduction

Gas chromatography (GC) is a commonly used chromatography in analytic chemistry for separating compounds that are gaseous or can be vaporized without decomposition. Because of its simplicity, sensitivity, and effectiveness in separating components of mixtures, it is widely used for quantitative and qualitative analysis of mixtures, for the purification of compounds, and for the determination of such thermochemical constants as heats of solution and vaporization, vapor pressure, and activity coefficients.

History

Archer J.P. Martin (Figure 1.1) and Anthony T. James (Figure 1.2) introduced liquid gas partition chromatography in 1950 at the meeting of the Biochemical Society held in London, a few months before submitting three fundamental papers to the Biochemical Journal. It was this work that provided the foundation for the development of gas chromatography.

Figure 1.1: British chemist Archer J. P. Martin, FRS (1910 - 2002) shared the Nobel Prize in 1952 for partition chromatography.

In fact, Martin envisioned gas chromatography almost ten years before, while working with R. L. M. Synge (Figure 1.3) on partition chromatography.

Martin and Synge, who were awarded the chemistry Nobel prize in 1941, suggested that separation of volatile compounds could be achieved by using a vapor as the mobile phase instead of a liquid. Gas chromatography quickly gained general acceptance because it was introduced at the time when improved analytical controls were required in the petrochemical industries, and new techniques were needed in order to overcome the limitations of old laboratory methods. Nowadays, gas chromatography is a mature technique, widely used worldwide for the analysis of almost every type of organic compound, even those that are not volatile in their original state but can be converted to volatile derivatives.

Figure 1.2: British chemist Anthony T. James (1922 - 2006).

Figure 1.3: British biochemist Richard L. M. Synge, FRS (1914 - 1994) shared the Nobel Prize in 1952 for partition chromatography.

The chromatographic process

Gas chromatography is a separation technique in which the components of a sample partition between two phases:

- The stationary phase.
- The mobile gas phase.

According to the state of the stationary phase, gas chromatography can be classified in gas-solid chromatography (GSC), where the stationary phase is a solid, and gas-liquid chromatography (GLC) that uses a liquid as stationary phase. GLC is to a great extent more widely used than GSC.

During a GC separation, the sample is vaporized and carried by the mobile gas phase (i.e., the carrier gas) through the column. Separation of the different components is achieved based on their relative vapor pressure and affinities for the stationary phase. The affinity of a substance towards the stationary phase can be described in chemical terms as an equilibrium constant called the distribution constant K_c, also known as the partition coefficient,

$$K_c = [A]_s/[A]_m$$

where $[A]_s$ is the concentration of compound A in the stationary phase and $[A]_m$ is the concentration of compound A in the stationary phase.

The distribution constant (K_c) controls the movement of the different compounds through the column, therefore differences in the distribution constant allow for the chromatographic separation. K_c is temperature dependent, and also depends on the chemical nature of the stationary phase. Thus, temperature can be used as a way to improve the separation of different compounds through the column, or a different stationary phase.

A typical chromatogram

Figure 1.4 shows a chromatogram of the analysis of residual methanol in biodiesel, which is one of the required properties that must be measured to ensure the quality of the product at the time and place of delivery.

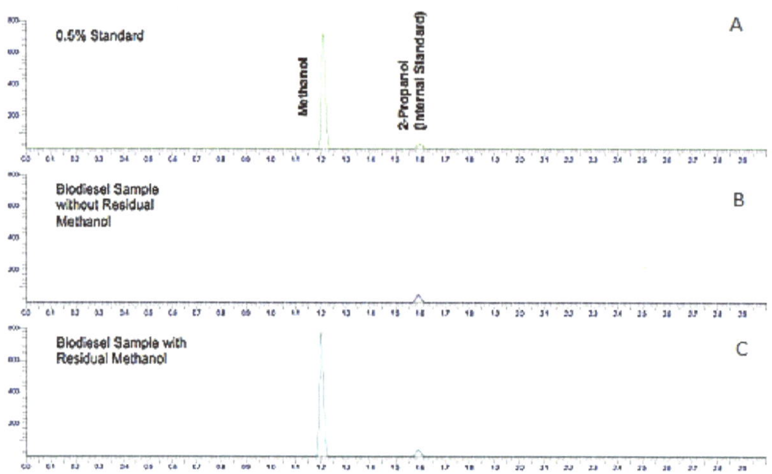

Figure 1.4: Chromatogram of the analysis of methanol in B100 biodiesel, following EN 14110 methodology. Reproduced courtesy of PerkinElmer Inc. (http://www.perkinelmer.com/).

Chromatogram (Figure 1.4a) shows a standard solution of methanol with 2-propanol (iso-propanol, IPA) as the internal standard. From the figure it can be seen that methanol has a higher affinity for the mobile phase (lower K_c) than 2-propanol, and therefore elutes first. Chromatograms (Figure 1.5b and c) show two samples of biodiesel, one with methanol (Figure 1.4b) and another with no methanol detection. The internal standard (IPA) was added to both samples for quantitation purposes.

Instrument overview

Components of a gas chromatograph system

Figure 1.5 shows a schematic diagram of the components of a typical gas chromatograph, while Figure 1.6 shows a photograph of a typical gas chromatograph coupled to a mass spectrometer (GC/MS).

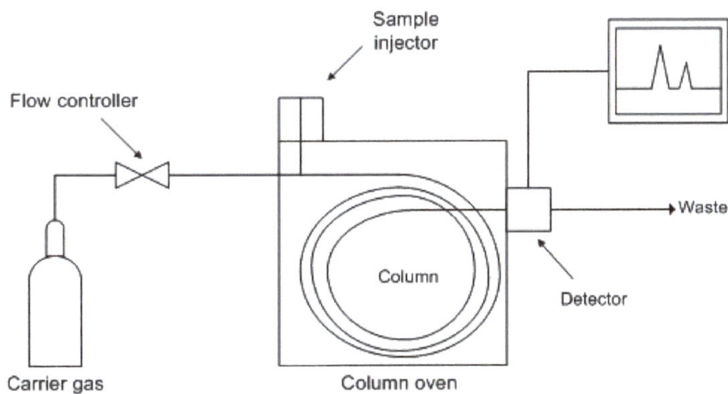

Figure 1.5: Schematic diagram of the components of a typical gas chromatograph.

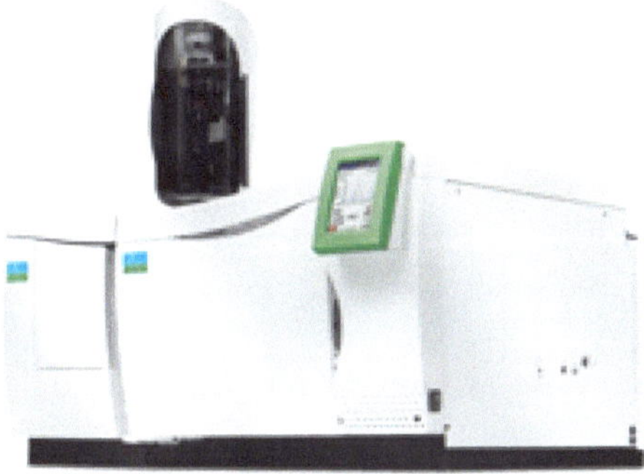

Figure 1.6: Image of a Perkin Elmer Clarus SQ 8S GC/MS. Reproduced courtesy of PerkinElmer Inc. (http://www.perkinelmer.com).

Carrier gas

The role of the carrier gas -GC mobile phase- is to carry the sample molecules along the column while they are not dissolved in or adsorbed on the stationary phase. The carrier gas is inert and does not interact with the sample, and thus GC separation's selectivity can be attributed to the stationary phase alone. However, the choice of carrier gas is important to maintain high efficiency. The effect of different carrier gases on column efficiency is represented by the van Deemter (packed columns) and the Golay equation (capillary columns). The van Deemter equation,

$$HEPT = A + (B/u) + (C \times u)$$

where, HEPT is the height equivalent to a theoretical plate (m), A is the eddy diffusion parameter (m), B is the diffusion coefficient (m^2/s), C is the resistance to mass transfer coefficient of the analyte between mobile and stationary phase (s), and u is the linear velocity (m/s). The three terms describe the three main effects that contribute to band broadening in packed columns and, as a consequence, to a reduced efficiency in the separation process.

These three factors are:
- the eddy diffusion (the A-term), which results from the fact that in packed columns spaces between particles along the column are not uniform. Therefore, some molecules take longer pathways than others, and there are also variations in the velocity of the mobile phase,
- the longitudinal molecular diffusion (the B-term) which is a consequence of having regions with different analyte concentrations,
- the mass transfer in the stationary liquid phase (the C-term)

The broadening is described in terms of the height equivalent to a theoretical plate, HEPT, as a function of the average linear gas velocity, u. A small HEPT value indicates a narrow peak and a higher efficiency.

Since capillary columns do not have any packing, the Golay equation,

$$HEPT = (B/u) + (C_s + C_m)u$$

does not have an A-term, but instead has two C-terms: one for mass transfer in then stationary phase (C_s) and one for mass transfer in the mobile phase (C_m).

High purity hydrogen, helium and nitrogen are commonly used for gas chromatography. Also, depending on the type of detector used, different gases are preferred.

Auto sampling system

An auto sampling system consists of auto sampler, and vaporization chamber. The sample to be analyzed is loaded at the injection port via a hypodermic syringe and it will be volatilized as the injection port is heated up. Typically, samples of one micro liter or less are injected on the column. These volumes

can be further reduced by using what is called a split injection system in which a controlled fraction of the injected sample is carried away by a gas stream before entering the column.

Injector

This is the place where the sample is volatilized and quantitatively introduced into the carrier gas stream. Usually a syringe is used for injecting the sample into the injection port. Samples can be injected manually or automatically with mechanical devices that are often placed on top of the gas chromatograph: the auto- samplers.

Column

The gas chromatographic column may be considered the heart of the GC system, where the separation of sample components takes place. Columns are classified as either packed or capillary columns. A general comparison of packed and capillary columns is shown in Table 1.1. Images of packed columns are shown in Figures 1.7 and 1.8.

Column type	Packed column	Capillary column
History	First type of GC column used	Modern technology. Today most GC applications are developed using capillary columns
Composition	Packed with silica particles onto which the stationary phase is coated.	Not packed with particulate material. Made of chemically treated fused silica covered with thin, uniform liquid phase films
Efficiency	Low	High
Outside diameter	2-4 mm	0.4 mm
Column length	2-4 meters	15-60 meters
Advantages	Lower cost, larger samples	Faster, better for complex mixtures

Table 1.1: A summary of the differences between a packed and a capillary column.

Since most common applications employed nowadays use capillary columns, we will focus on this type of columns. To define a capillary column, four parameters must be specified:

The stationary phase is the parameter that will determine the final resolution obtained and will influence other selection parameters. Changing the stationary phase is the most powerful way to alter selectivity in GC analysis.

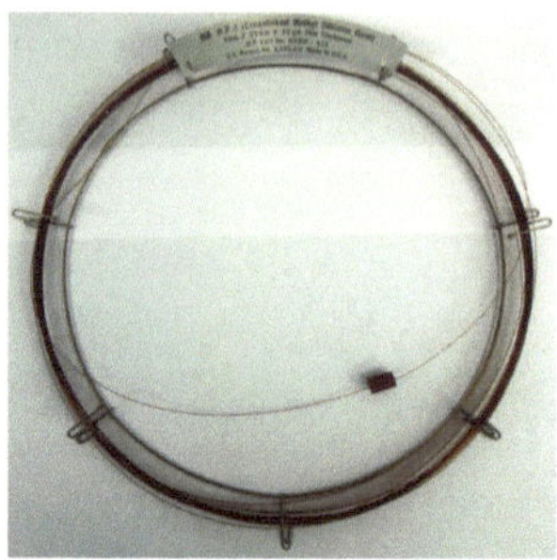

Figure 1.7: A typical capillary GC column.

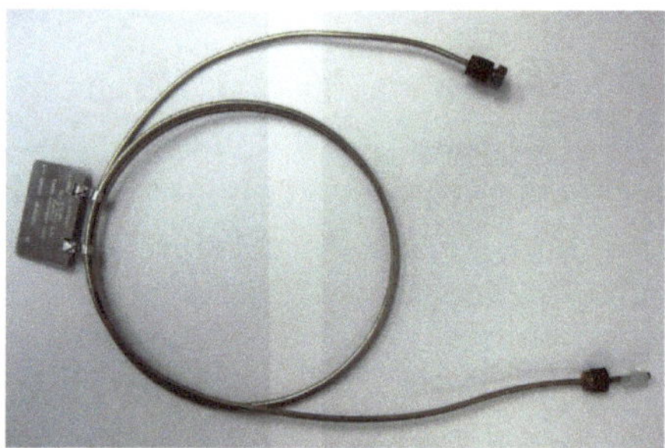

Figure 1.8: Silica packed GC columns.

The length is related to the overall efficiency of the column and to overall analysis time. A longer column will increase the peak efficiency and the quality of the separation, but it will also increase analysis time. One of the classical trade-offs in gas chromatography (GC) separations lies between speed of analysis and peak resolution.

The column internal diameter (ID) can influence column efficiency (and therefore resolution) and also column capacity. By decreasing the column internal diameter, better separations can be achieved, but column overload and peak broadening may become an issue.

The sample capacity of the column will also depend on film thickness. Moreover, the retention of sample components will be affected by the thickness of the film, and therefore its retention time. A shorter run time and higher resolution can be achieved using thin films; however, these films offer lower capacity.

Detector

The detector senses a physicochemical property of the analyte and provides a response which is amplified and converted into an electronic signal to produce a chromatogram. Most of the detectors used in GC were invented specifically for this technique, except for the thermal conductivity detector (TCD) and the mass spectrometer. In total, approximately 60 detectors have been used in GC. Detectors that exhibit an enhanced response to certain analyte types are known as "selective detectors".

During the last 10 years there had been an increasing use of GC in combination with mass spectrometry (MS). The mass spectrometer has become a standard detector that allows for lower detection limits and does not require the separation of all components present in the sample. Mass spectroscopy is one of the types of detection that provides the most information with only micrograms of sample. Qualitative identification of unknown compounds as well as quantitative analysis of samples is possible using GC-MS. When GC is coupled to a mass spectrometer, the compounds that elute from the GC column are ionized by using electrons (EI, electron ionization) or a chemical reagent (CI, chemical ionization). Charged fragments are focused and accelerated into a mass analyzer: typically, a quadrupole mass analyzer. Fragments with different mass to charge ratios will generate different signals, so any compound that produces ions within the mass range of the mass analyzer will be detected. Detection limits of 1-10 ng or even lower values (e.g., 10 pg) can be achieved selecting the appropriate scanning mode.

Recording devices

GC system originally used paper chart readers, but modern system typically uses an online computer, which can track and record the electrical signals of the separated peaks. The data can be later analyzed by software to provide the information of the gas mixture.

Separation terminology

An ideal separation is judged by resolution, efficiency, and symmetry of the desired peaks, as illustrated by Figure 1.9.

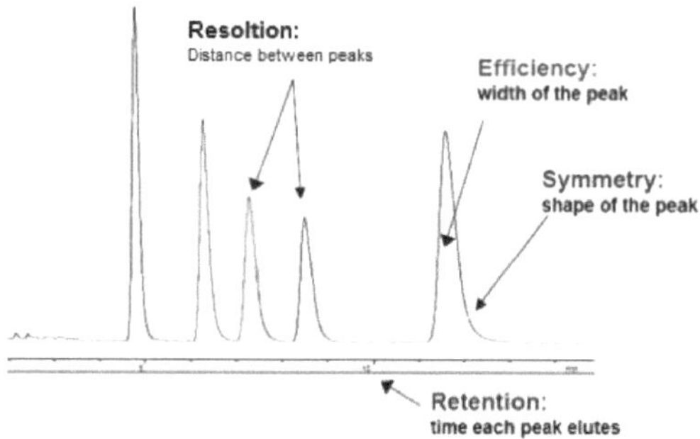

Figure 1.9: Separation terminology.

Resolution (R)

Resolution can be simply expressed as the distance on the output trace between two peaks. The highest possible resolution is the goal when developing a separation method. Resolution is defined by the R value,

$$R = \text{capacity} \times \text{selectivity} \times \text{efficiency}$$

which can be expressed mathamatically,

$$R = |k/(1+k)|(\alpha - 1/\alpha)(N^{0.5}/4)$$

where k is capacity, α is selectivity, and N is the number of theoretical plates. An R value of 1.5 is defined as being the minimum required for baseline separation, i.e., the two adjacent peaks are separated by the baseline. Separation for different R values is illustrated in Figure 1.10.

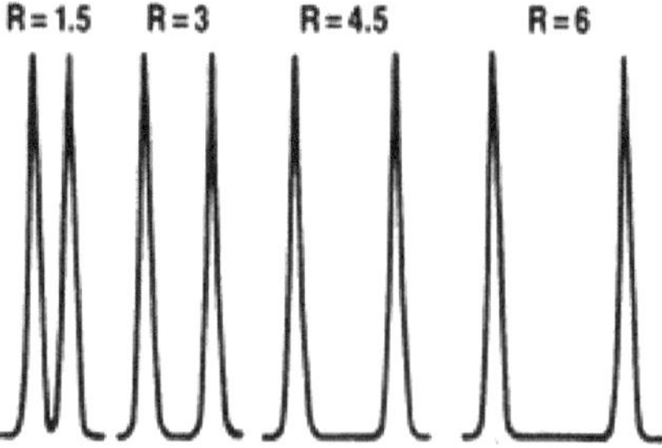

Figure 1.10: Different separation resolutions.

Capacity (k')

Capacity (k') is known as the retention factor. It is a measure of retention by the stationary phase. It is calculated from,

$$k' = (t_r - t_m)/t_m$$

where t_r = retention time of analyte (substance to be analyzed), and t_m = retention time of an unretained compound.

Selectivity

Selectivity is related to α, the separation factor (Figure 1.11). The value of α should be large enough to give baseline resolution but minimized to prevent waste.

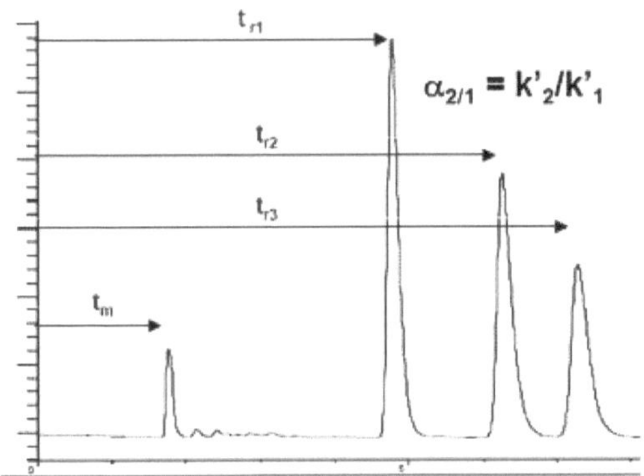

Figure 1.11: Scheme for the calculation of selectivity.

Efficiency

Narrow peaks have high efficiency (Figure 1.12) and are desired. Units of efficiency are "theoretical plates" (N) and are often used to describe column performance. "Plates" is the current common term for N, is defined as a function of the retention time (t_r) and the full peak width at half maximum ($W_{b1/2}$),

$$N = 5.545 (t_r/W_{b1/2})^2$$

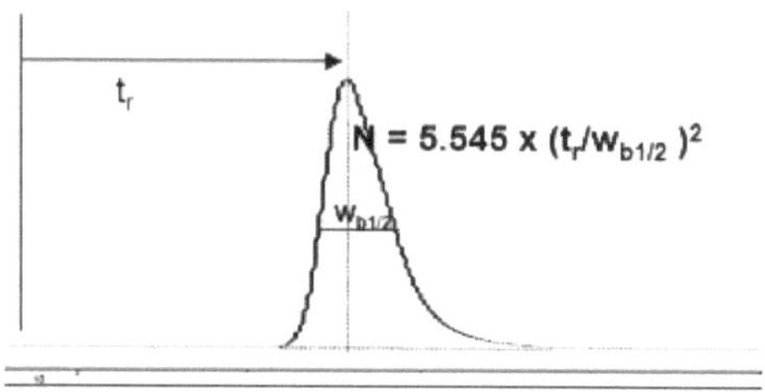

Figure 1.12: Scheme for calculating efficiency.

Peak symmetry

The symmetry of a peak is judged by the values of two half peak widths, a and b (Figure 1.13). When a = b, a peak is called symmetric, which is desired. Unsymmetrical peaks are often described as "tailing" or "fronting".

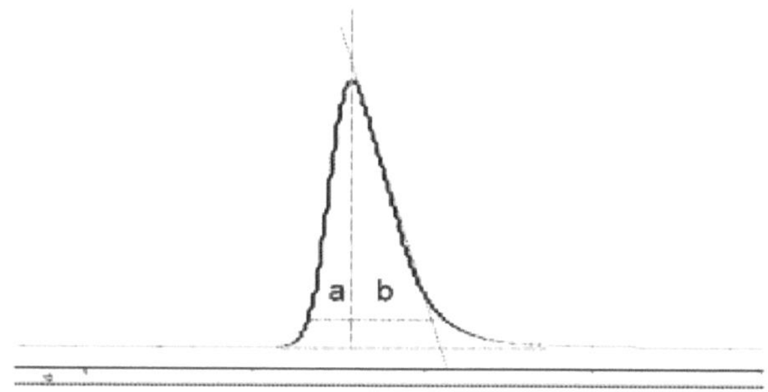

Figure 1.13: Scheme for the symmetry of a peak.

An ideal separation

The attributions of an ideal separation are as follows:
- Should meet baseline resolution of the compounds of interest. Each desired peak is narrow and symmetrical.
- Has no wasted dead time between peaks.
- Takes a minimal amount of time to run.
- The result is reproducible.

Choosing a method

Carrier gas and flow rate

Helium, nitrogen, argon, hydrogen and air are typically used carrier gases. Which one is used is usually determined by the detector being used, for example, a discharge ionization detection (DID) requires helium as the carrier gas. When analyzing gas samples, however, the carrier is sometimes selected based on the sample's matrix, for example, when analyzing a mixture in argon, an argon carrier is preferred, because the argon in the sample does not show up on the chromatogram. Safety and availability are other factors, for example, hydrogen is flammable, and high-purity helium can be di cult to obtain in some areas of the world.

The carrier gas flow rate affects the analysis in the same way that temperature does. The higher the flow rate the faster the analysis, but the lower the separation between analytes. Furthermore, the shape of peak will be also affected by the flow rate. The slower the rate is, the more axial and radical diffusion are, the broader and the more asymmetric the peak is. Selecting the flow rate is therefore the same compromise between the level of separation and length of analysis as selecting the column temperature.

Column Selection

Table 1.2 and Figure 1.14 show commonly used stationary phase in various applications.

Column temperature and temperature program

For precise work, the column temperature must be controlled to within tenths of a degree. The optimum column temperature is dependent upon the boiling point of the sample. As a rule of thumb, a temperature slightly above the average boiling point of the sample results in an elution time of 2 - 30 minutes. Minimal temperatures give good resolution but increase elution times. If a sample has a wide boiling range, then temperature programming can be useful. The column temperature is increased (either continuously or in steps) as separation proceeds. Another effect that temperature may have is on the shape of peak as flow rate does. The higher the temperature is, the more intensive the diffusion is, the worse the shape is. Thus, a compromise has to be made between goodness of separation and retention time as well as peak shape.

Stationary phase	Common tradename	Maximum temp. (°C)	Applications
Polydimethyl siloxane	OV-1, SE-30	350	General-purpose non-polar phase, hydrocarbons, polynuclear aromatics, steroids, PCBs drugs
Poly(phenylmethyl-dimethyl) siloxane (10% phenyl)	OV-3, SE-52	350	Fatty acid methyl esters, alkaloids, drugs, halogenated compounds
Poly(phenyl methyl) siloxane (50% phenyl)	OV-17	250	Drugs, steroids, pesticides, glycols
Poly(trifluoro propyl methyl) siloxane	OV-210	200	Chlorinated aromatics, nitroaromatics, alkyl- substituted benzenes
Polyethylene glycol	Carbowax 20M	250	Free acids, alcohols, ethers, essential oils, glycols
Poly(dicyano propyl) siloxane	OV-275	240	Polyunsaturated fatty acid, rosin acids, free acids, alcohols

Table 1.2: Some common stationary phases for gas chromatography. The structures of which are shown in Figure 1.14.

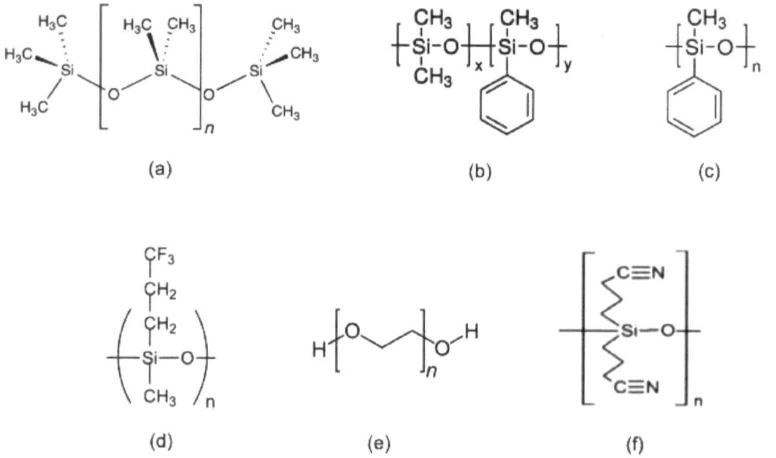

Figure 1.14: Structure of stationary phases for gas chromatography: (a) polydimethyl siloxane, (b) poly(phenylmethyl-dimethyl) siloxane, (c) poly(phenylmethyl) siloxane, (d) poly(trifluoropropyl-dimethyl) siloxane, (e) polyethylene glycol, and (f) Poly(dicyanoallyldimethyl) siloxane

Detector selection

A number of detectors are used in gas chromatography. The most common are the flame ionization detector (FID) and the thermal conductivity detector (TCD). Both are sensitive to a wide range of components, and both work over a wide range of concentrations. While TCDs are essentially universal and can be used to detect any component other than the carrier gas (as long as their thermal conductivities are different from that of the carrier gas, at detector temperature), FIDs are sensitive primarily to hydrocarbons, and are more sensitive to them than TCD. However, an FID cannot detect water. Both detectors are also quite robust. Since TCD is non-destructive, it can be operated in-series before an FID (destructive), thus providing complementary detection of the same analytes. ECD is a very sensitive detection, for halides, nitrates, nitriles, peroxides, anhydrides and organometallics, which can detect up to 50 fg (i.e., 50×10^{-15} g) of those analytes. Different types of detectors are listed below in Table 1.3, along with their properties.

Detector	Support gases	Selectivity	Detectability	Dynamic range
Flame ionization (FID)	Mass flow	Most organic compounds	100 pg	10^7
Thermal conductivity (TCD)	Reference	Universal	1 ng	10^7
Electron capture (ECD)	Make-up	Halides, nitrates, nitriles, peroxides, anhydrides, organometallics	50 fg	10^5
Nitrogen- phosphorus	Hydrogen and air	Nitrogen, phosphorus	10 pg	10^6
Flame photometric (FPD)	Hydrogen and air possibly oxygen	Sulphur, phosphorus, tin, boron, arsenic, germanium, selenium, chromium	100 pg	10^3
Photo-ionization (PID)	Make-up	Aliphatics, aromatics, ketones, esters, aldehydes, amines, heterocyclics, organo-sulfurs	2 pg	10^7
Hall electrolytic conductivity	Hydrogen, oxygen	Halide, nitrogen, nitrosamine, sulfur	-	-

Table 1.3: Different types of detectors and their properties.

Sample preparation techniques

Gas chromatography is primarily used for the analysis of thermally stable volatile compounds. However, when dealing with non-volatile samples, chemical reactions can be performed on the sample to increase the volatility of the compounds. Compounds that contain functional groups such as OH, NH, CO_2H, and SH are difficult to analyze by GC because they are not sufficiently volatile, can be too strongly attracted to the stationary phase or are thermally unstable. Most common derivatization reactions used for GC can be divided into three types:

- silylation,
- acylation,
- alkylation and esterification.

Samples are derivatized before being analyzed to:
- Increase volatility and decrease polarity of the compound,
- Reduce thermal degradation,
- Increase sensitivity by incorporating functional groups that lead to higher detector signals,
- Improve separation and reduce tailing.

Advantages and disadvantages

GC is the premier analytical technique for the separation of volatile compounds. Several features such as speed of analysis, ease of operation, excellent quantitative results, and moderate costs had helped GC to become one of the most popular techniques worldwide.

Advantages of GC
- Due to its high efficiency, GC allows the separation of the components of complex mixtures in a reasonable time.
- Accurate quantitation (usually sharp reproducible peaks are obtained)
- Mature technique with many applications notes available for users.
- Multiple detectors with high sensitivity (ppb) are available, which can also be used in series with a mass spectrometer since MS is a non-destructive technique.

Disadvantages of GC
- Limited to thermally stable and volatile compounds.

- Most GC detectors are destructive, except for MS.

Gas chromatography versus high performance liquid chromatography (HPLC)

Unlike gas chromatography, which is unsuitable for nonvolatile and thermally fragile molecules, liquid chromatography can safely separate a very wide range of organic compounds, from small-molecule drug metabolites to peptides and proteins (Table 1.4).

GC	HPLC
Sample must be volatile or derivatized previous to GC analysis	Volatility is not important, however solubility in the mobile phase becomes critical for the analysis.
Most analytes have a molecular weight (Mw) below 500 Da (due to volatility issues)	There is no upper molecular weight limit as far as the sample can be dissolved in the appropriate mobile phase
Can be coupled to MS. Several mass spectral libraries are available if using electron ionization	Methods must be adapted before using an MS detector (non-volatile buffers cannot be used)
Can be coupled to several detectors depending on the application	For some detectors the solvent must be an issue. When changing detectors some methods will require prior modification

Table 1.4: Relative advantages and disadvantages of GC versus HPLC.

Headspace analysis using GC

Most consumer products and biological samples are composed of a wide variety of compounds that differ in molecular weight, polarity, and volatility. For complex samples like these, headspace sampling is the fastest and cleanest method for analyzing volatile organic compounds. A headspace sample is normally prepared in a vial containing the sample, the dilution solvent, a matrix modifier, and the headspace (Figure 1.15). Volatile components from complex sample mixtures can be extracted from non-volatile sample components and isolated in the headspace or vapor portion of a sample vial. An aliquot of the vapor in the headspace is delivered to a GC system for separation of all of the volatile components.

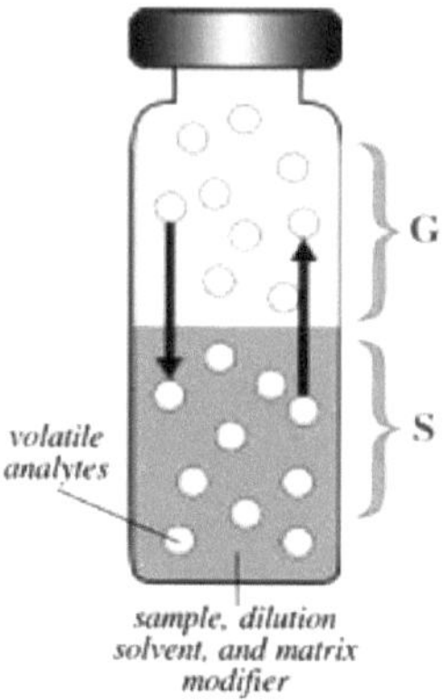

Figure 1.15: Schematic representation of the phases of the headspace in the vial. Adapted from *A Technical Guide for Static Headspace Analysis Using GC*, Restek Corp. (2000).

The gas phase (G in Figure 1.15) is commonly referred to as the headspace and lies above the condensed sample phase. The sample phase (S in Figure 1.15) contains the compound(s) of interest and is usually in the form of a liquid or solid in combination with a dilution solvent or a matrix modifier. Once the sample phase is introduced into the vial and the vial is sealed, volatile components di use into the gas phase until the headspace has reached a state of equilibrium as depicted by the arrows. The sample is then taken from the headspace.

Basic principles of headspace analysis

Partition coefficient

Samples must be prepared to maximize the concentration of the volatile components in the headspace and minimize unwanted contamination from other compounds in the sample matrix. To help determine the concentration of an analyte in the headspace, you will need to calculate the partition coefficient (K), which is defined by,

$$K = C_s/C_g$$

where C_s is the concentration of analyte in sample phase and C_g is the concentration of analyte in gas phase. Compounds that have low K values will tend to partition more readily into the gas phase and have relatively high responses and low limits of detection. K can be lowered by changing the temperature at which the vial is equilibrated or by changing the composition of the sample matrix.

Phase ratio

The phase ratio (β) is defined as the relative volume of the headspace compared to volume of the sample in the sample vial,

$$\beta = V_g/V_s$$

where V_s = volume of sample phase and V_g = volume of gas phase. Lower values for β (i.e., larger sample size) will yield higher responses for volatile compounds. However, decreasing the β value will not always yield the increase in response needed to improve sensitivity. When β is decreased by increasing the sample size, compounds with high K values partition less into the headspace compared to compounds with low K values and yield correspondingly smaller changes in C_g =. Samples that contain compounds with high K values need to be optimized to provide the lowest K value before changes are made in the phase ratio.

GC analysis of the hydrodechlorination reaction of trichloroethene

Trichloroethene (TCE) is a widely spread environmental contaminant and a member of the class of compounds known as dense non-aqueous phase liquids (DNAPLs). Pd/Al_2O_3 catalyst has shown activity for the hydrodechlorination (HDC) of chlorinated compounds.

To quantify the reaction rate, a 250 mL screw-cap bottle with 77 mL of headspace gas was used as the batch reactor for the studies. TCE (3 µL) is added in 173 mL DI water purged with hydrogen gas for 15 mins, together with 0.2 µL pentane as internal standard. Dynamic headspace analysis using GC has been applied. The experimental condition is concluded in the table below (Table 1.5).

TCE	3 µL
H_2	1.5 ppm
pentane	0.2 µL
DI water	173 mL
1 wt% Pd/Al_2O_3	50 mg
Temperature	25 °C
Pressure	1 atm
Reaction time	1h

Table 1.5: The experimental condition in HDC of TCE.

Reaction kinetics

First order reaction is assumed in the HDC of TCE,

$$-dC_{TCE}/dt = k_{meas} \times C_{TCE}$$

where K_{means} is defined by,

$$k_{meas} = k_{cat} \times C_{cat}$$

and C_{cat} is equal to the concentration of Pd metal within the reactor and k_{cat} is the reaction rate with units of $L/g_{Pd}/min$.

The GC parameters

The GC methods used are listed in Table 1.6.

GC type	Agilent 6890N GC
Column	Supelco 1-2382 40/60 Carboxen-1000 packed column
Detector	FID
Oven temperature	210 °C
Flow rate	35 mL/min
Injection amount	200 µL
Carrier gas	Helium
Detection time	5 min

Table 1.6: GC method for detection of TCE and other related chlorinated compounds.

Quantitative method

Since pentane (C_5H_{12}) is introduced as the inert internal standard, the relative concentration of TCE in the system can be expressed as the ratio of area of TCE to pentane in the GC plot,

$$C_{TCE} = \text{(peak area of TCE)/(peak area of pentane)}$$

Data analysis

The major analytes (referenced as TCE, pentane, and ethane) are very well separated from each other, allowing for quantitative analysis. The peak areas of the peaks associated with these compounds are integrated by the computer automatically and are listed in (Table 1.7) with respect to time.

Time/min	Peak area of pentane	Peak area of TCE
0	5992.93	13464
5.92	6118.5	11591
11.25	5941.2	8891
16.92	5873.5	7055.6
24.13	5808.6	5247.4
32.65	5805.3	3726.3
43.65	5949.8	2432.8
53.53	5567.5	1492.3
64.72	5725.6	990.2
77.38	5624.3	550
94.13	5432.5	225.7
105	5274.4	176.8

Table 1.7: Peak area of pentane, TCE as a function of reaction time.

Normalize TCE concentration with respect to peak area of pentane and then to the initial TCE con- centration, and then calculate the nature logarithm of this normalized concentration, as shown in Table 1.8.

From a plot normalized TCE concentration against time shows the concentration profile of TCE during reaction (Figure 1.16), while the slope of the logarithmic plot provides the reaction rate constant (Figure 1.17).

Time (min)	TCE/pentane	TCE/pentane/$TCE_{initial}$	ln(TCE/Pentane/$TCE_{initial}$)
0	2.2466	1.0000	0.0000
5.92	1.8944	0.8432	-0.1705
11.25	1.4965	0.6661	-0.4063
16.92	1.2013	0.5347	-0.6261
24.13	0.9034	0.4021	-0.9110
32.65	0.6419	0.2857	-1.2528
43.65	0.4089	0.1820	-1.7038
53.53	0.2680	0.1193	-2.1261
64.72	0.1729	0.0770	-2.5642
77.38	0.0978	0.0435	-3.1344
94.13	0.0415	0.0185	-3.9904
105	0.0335	0.0149	-4.2050

Table 1.8: Normalized TCE concentration as a function of reaction time.

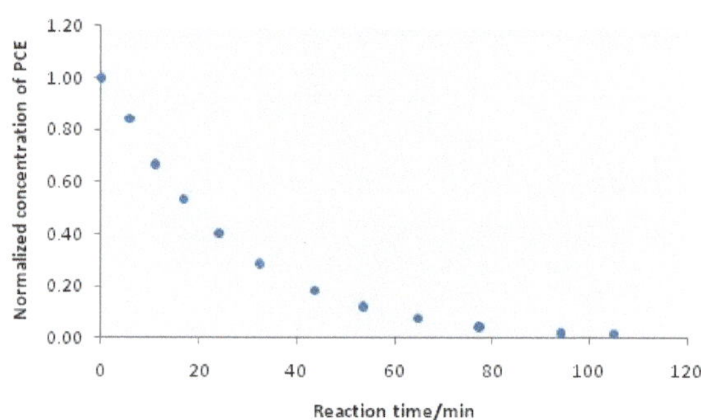

Figure 1.16: A plot of the normalized concentration profile of TCE.

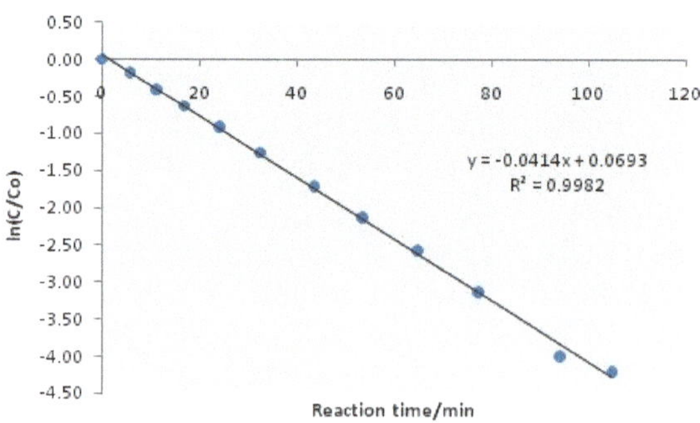

Figure 1.17: A plot of ln(C_{TCE}/C_0) versus time.

From Figure 1.17, we can see that the linearity, i.e., the goodness of the assumption of first order reaction, is very much satisfied throughout the reaction. Thus, the reaction kinetic model is validated. Furthermore, the reaction rate constant can be calculated from the slope of the fitted line, i.e., k_{meas} = 0.0414 min^{-1}. From this the k_{cat} can be obtained:

$$k_{cat} = k_{meas}/C_{Pd} = \frac{0.0414 \text{ min}^{-1}}{(5 \times 10^{-4} \text{ g}/0.173 \text{ L})} = 14.32 \text{L/g}_{Pd} \text{ min}$$

Bibliography

A Technical Guide for Static Headspace Analysis Using GC, Restek Corp. (2000).

E. F. Barry, *Columns for Gas Chromatography: Performance and Selection*, Wiley-Interscience, Hoboken, NJ (2007)

L. S. Ettre, Evolution of capillary columns for gas chromatography. *LCGC*, 2001, **19**, 48.

R. L. Grob and E. F. Barry, *Modern Practice of Gas Chromatography*, 4th edn., Wiley-Interscience, Hoboken, NJ (2004).

D. C. Harris, *Quantitative Chemical Analysis*, 5th edn., Freeman and Company (1999).

J. V. Hinshaw, Practical gas chromatography. *LCGC North America*, 2013, **31**, 932.

A.T. James, Gas-liquid partition chromatography: the separation of volatile aliphatic amines and of the homologues of pyridine. *Biochem. J.*, 1952, **52**, 242.

A. T. James and A. J. P. Martin, Gas-liquid partition chromatography: the separation and micro-estimation of volatile fatty acids from formic acid to dodecanoic acid. *Biochem. J.*, 1952, **50**, 679.

A. T. James, A. J. P. Martin, and G. H. Smith, Gas-liquid partition chromatography: the separation and micro-estimation of ammonia and the methylamines. *Biochem. J.*, 1952, **52**, 238.

S. J. Maguire-Boyle and A. R. Barron, Organic compounds in produced waters from shale gas wells. *Environ. Sci.: Processes Impacts*, 2014, **16**, 2237.

A. J. P. Martin and R. L. M Synge, A new form of chromatogram employing two liquid phases: A theory of chromatography. 2. Application to the micro-determination of the higher monoamino-acids in proteins. *Biochem. J.*, 1941, **35**, 1358.

G. McMahon, *Analytical Instrumentation: A Guide to Laboratory, Portable and Miniaturized Instruments*, 1st edn., Wiley, Hoboken, N.J (2007)

H. M. McNair, *Basic Gas Chromatography*, Wiley, New York (1998).

M. O. Nutt, J, B. Hughes, and M, S. Wong, Designing Pd-on-Au bimetallic nanoparticle catalysts for trichloroethene hydrodechlorination. *Environ. Sci. Technol.*, 2005, **39**, 1346.

D. L, Pavia, G. M. Lampman, G. S. Kritz, and R. G. Engel, *Introduction to Organic Laboratory Techniques*, 4th edn., Thomson Brooks/Cole (2006).

Chapter 2: High Performance Liquid Chromatography

Sha Li and Andrew R. Barron

Introduction

High-performance liquid chromatography (HPLC) is a technique in analytical chemistry used to separate the components in a mixture, and to identify and quantify each component. It was initially discovered as an analytical technique in the early twentieth century and was first used to separate colored compounds. The word chromatography means color writing. It was the botanist M. S. Tswett (Figure 2.1) who invented this method in around 1900 to study leaf pigments, mainly chlorophyll (Figure 2.2). He separated the pigments based on their interaction with a stationary phase.

Figure 2.1: Russian born Italian botanist Mikhail Semyonovich Tswett (1872 - 1919).

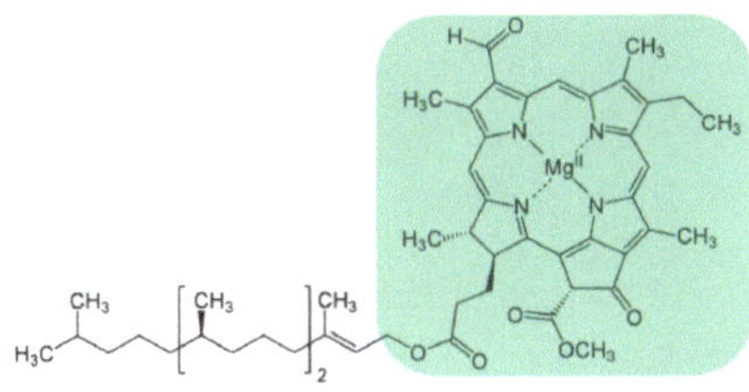

Figure 2.2: Structure of one of the many types of chlorophyll, highlighting the chlorin magnesium unit that is common to all chlorophylls.

In 1906 Tswett published two fundamental papers describing the various aspects of liquid-adsorption chromatography in detail. He also pointed out that in spite of its name, other substances also could be separated by chromatography. The modern high-performance liquid chromatography has developed from this separation; the separation efficiency, versatility and speed have been improved significantly.

The molecular species subjected to separation exist in a sample that is made of analytes and matrix. The analytes are the molecular species of interest, and the matrix is the rest of the components in the sample. For chromatographic separation, the sample is introduced in a flowing mobile phase that passes a stationary phase. Mobile phase is a moving liquid, and is characterized by its composition, solubility, UV transparency, viscosity, and miscibility with other solvents. Stationary phase is a stationary medium, which can be a stagnant bulk liquid, a liquid layer on the solid phase, or an interfacial layer between liquid and solid. In HPLC, the stationary phase is typically in the form of a column packed with very small porous particles and the liquid mobile phase is moved through the column by a pump. The development of HPLC is mainly the development of the new columns, which requires new particles, new stationary phases (particle coatings), and improved procedures for packing the column. A picture of modern HPLC is shown in Figure 2.3.

Figure 2.3: A picture of modern HPLC instrument.

Instrumentation

The major components of a HPLC are shown in Figure 2.4. The role of a pump is to force a liquid (mobile phase) through at a specific flow rate (milliliters per minute). The injector serves to introduce the liquid sample into the flow stream of the mobile phase. Column is the most central and important component of HPLC, and the column's stationary phase separates the sample components of interest using various physical and chemical parameters. The detector is to detect the individual molecules that elute from the column. The computer usually functions as the data system, and the computer not only controls all the modules of the HPLC instrument but it takes the signal from the detector and uses it to determine the retention time, the sample components, and quantitative analysis.

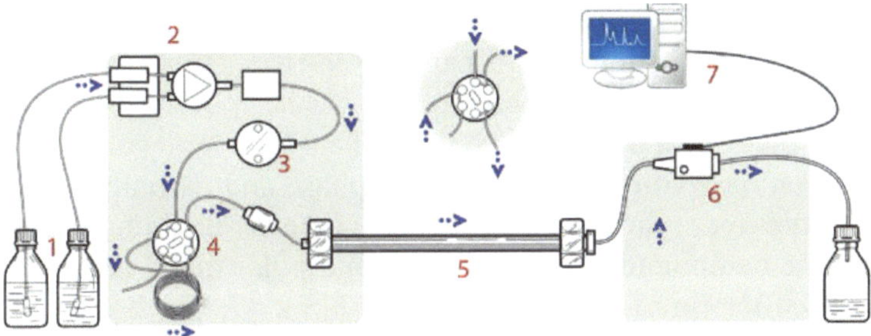

Figure 2.4: Schematic representation of a HPLC system: (1) solvent, (2) gradient valve, (3) high- pressure pump, (4) sample injection loop, (5) analytical column, (6) detector, and (7) computer.

Columns

Different separation mechanisms were used based on different property of the stationary phase of the column. The major types include normal phase chromatography, reverse phase chromatography, ion exchange, size exclusion chromatography, and affinity chromatography.

Normal-phase chromatography

In this method the columns are packed with polar, inorganic particles and a nonpolar mobile phase is used to run through the stationary phase (Table 2.1). Normal phase chromatography is mainly used for purification of crude samples, separation of very polar samples, or analytical separations by thin layer chromatography. One problem when using this method is that, water is a strong solvent for the normal-phase chromatography, traces of water in the mobile phase can markedly affect sample retention, and after changing the mobile phase, the column equilibration is very slow.

	Stationary phase	Mobile phase
Normal phase	Polar	Non polar
Reverse phase	Non polar	Polar

Table 2.1: Mobile phase and stationary phase used for normal phase and reverse-phase chromatography

Reverse-phase chromatography

In reverse-phase (RP) chromatography the stationary phase has a hydrophobic character, while the mobile phase has a polar character. This is the reverse of the normal-phase chromatography (Table 2.1). The interactions in RP-HPLC are considered to be the hydrophobic forces, and these forces are caused by the energies resulting from the disturbance of the dipolar structure of the solvent. The separation is typically based on the partition of the analyte between the stationary phase and the mobile phase. The solute molecules are in equilibrium between the hydrophobic stationary phase and partially polar mobile phase. The more hydrophobic molecule has a longer retention time while the ionized organic compounds, inorganic ions and polar metal molecules show little or no retention time.

Ion exchange chromatography

The ion exchange mechanism is based on electrostatic interactions between hydrated ions from a sample and oppositely charged functional groups on the stationary phase. Two types of mechanisms are used for the separation: in one mechanism, the elution uses a mobile phase that contains competing ions that would replace the analyte ions and push them off the column; another mechanism is to add a complexing reagent in the mobile phase and to change the sample species from their initial form. This modification on the molecules will lead them to elution. In addition to the exchange of ions, ion-exchange stationary phases are able to retain specific neutral molecules. This process is related to the retention based on the formation of complexes, and specific ions such as transition metals can be retained on a cation-exchange resin and can still accept lone-pair electrons from donor ligands. Thus, neutral ligand molecules can be retained on resins treated with the transitional metal ions.

The modern ion exchange is capable of quantitative applications at rather low solute concentrations and can be used in the analysis of aqueous samples for common inorganic anions (range 10 µg/L to 10 mg/L). Metal cations and in-organic anions are all separated predominantly by ionic interactions with the ion exchange resin. One of the largest industrial users of ion exchange is the food and beverage sector to determine the nitrogen-, sulfur-, and phosphorous- containing species as well as the halide ions. Also, ion exchange can be used to determine the dissolved inorganic and organic ions in natural and treated waters.

Size exclusion chromatography

It is a chromatographic method that separate the molecules in the solutions based on the size (hydrodynamic volume). This column is often used for the separation of macromolecules and of macromolecules from small molecules. After the analyte is injected into the column, molecules smaller than he pore size of the stationary phase enter the porous particles during the separation and flow through the intricate channels of the stationary phase. Thus, smaller components have a longer path to traverse and elute from the column later than the larger ones. Since the molecular volume is related to molecular weight, it is expected that retention volume will depend to some degree on the molecular weight of the polymeric materials. The relation between the retention time and the molecular weight is shown in Figure 2.5.

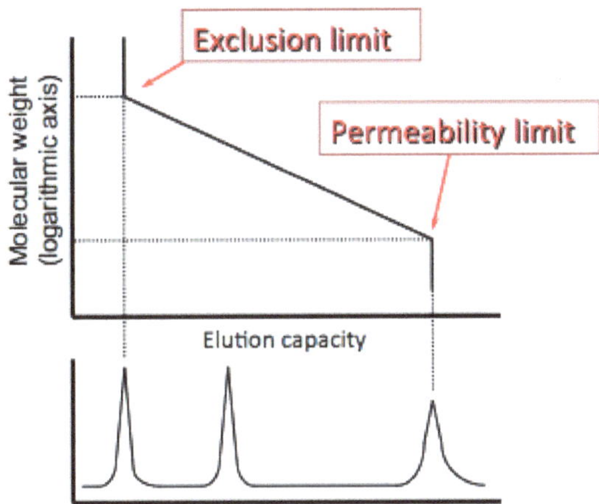

Figure 2.5: Graph showing the relationship between the retention time and molecular weight in size exclusion chromatography.

Usually the type of HPLC separation method to use depends on the chemical nature and physicochemical parameters of the samples. Figure 2.6 shows a flow chart of preliminary selection for the separation method according to the properties of the analyte.

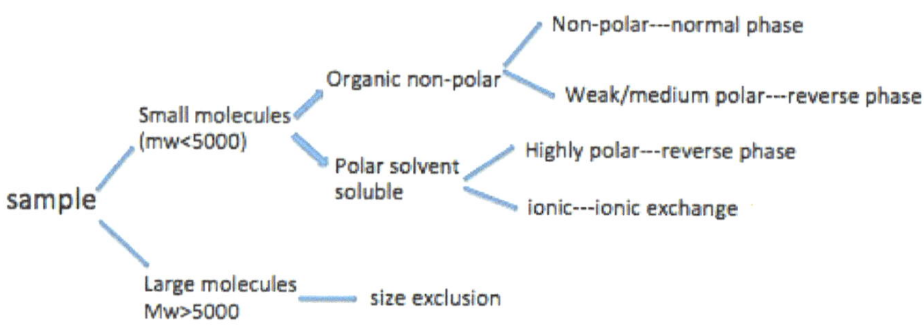

Figure 2.6: Diagram showing the sample properties related to the selection of HPLC type of analysis.

Detectors

Detectors that are commonly used for liquid chromatography include ultraviolet-visible absorbance detectors, refractive index detectors, fluorescence detectors, and mass spectrometry. Regardless of the class, a LC detector should ideally have the characteristics of about 10^{-12} -10^{-11} g/mL, and a linear dynamic range of five or six orders. The principal characteristics of the detectors to be evaluated include dynamic range, response index or linearity, linear dynamic range, detector response, detector sensitivity, etc.

Among these detectors, the most economical and popular methods are UV and refractive index (RI) detectors. They have rather broad selectivity reasonable detection limits most of the time. The RI detector was the first detector available for commercial use. This method is particularly useful in the HPLC separation according to size, and the measurement is directly proportional to the concentration of polymer and practically independent of the molecular weight. The sensitivity of RI is 10^{-6} g/mL, the linear dynamic range is from 10^{-6} - 10^{-4} g/mL, and the response index is between 0.97 and 1.03.

UV detectors respond only to those substances that absorb UV light at the wavelength of the source light. A great many compounds absorb light in the UV range (180 - 350 nm) including substances having one or more double bonds and substances having unshared electrons. and the relationship between the intensity of UV light transmitted through the cell and solute concentration is given by Beer's law,

$$I_T = I_0\, e^{-kcl}$$

$$\ln I_T = \ln I_0\, (-kcl)$$

Where I_0 is the intensity of the light entering the cell, and I_T is the light transmitted through the cell, l is the path length of the cell, c is the concentration of the solute, and k is the molar absorption coefficient of the solute. UV detectors include fixed wavelength UV detector and multi wavelength UV detector. The fixed wavelength UV detector has sensitivity of 5×10^{-8} g/mL, has linear dynamic range between 5×10^{-8} - 5×10^{-4} g/mL, and the response index is between 0.98 - 1.02. The multi-wavelength UV detector has sensitivity of 10^{-7} g/mL, the linear dynamic range is between 5×10^{-7} - 5×10^{-4} g/mL, and the response index is from 0.97 - 1.03. UV detectors could be used effectively for the reverse-phase separations and ion exchange chromatography. UV detectors have high sensitivity, are economically affordable, and easy to operate. Thus, UV detector is the most common choice of detector for HPLC.

Another method, mass spectrometry, has certain advantages over other techniques. Mass spectra could be obtained rapidly; only small amount (sub-μg) of sample is required for analysis, and the data provided by the spectra is very informative of the molecular structure. Mass spectrometry also has strong advantages of specificity and sensitivity compared with other detectors. The combination of HPLC-MS is oriented towards the specific detection and potential identification of chemicals in the presence of other chemicals. However, it is di cult to interface the liquid chromatography to a mass-spectrometer, because all the solvents need to be removed first. The commonly used interface includes, electrospray ionization, atmospheric pressure photoionization, and thermospray ionization.

Parameters related to HPLC separation

Flow rate

Flow rate shows how fast the mobile phase travels across the column and is often used for calculation of the consumption of the mobile phase in a given time interval. There are volumetric flow rate U and linear flow rate u. These two flow rates are related by,

$$U = Au$$

where A is the area of the channel for the flow,

$$A = (1/4)\pi\varepsilon d^2$$

Retention time

The retention time (t_R) can be defined as the time from the injection of the sample to the time of compound elution, and it is taken at the apex of the peak that belongs to the specific molecular species. The retention time is decided by several factors including the structure of the specific molecule, the flow rate of the mobile phase, column dimension. And the dead time t_0 is defined as the time for a non-retained molecular species to elute from the column.

Retention volume

Retention volume (V_R) is defined as the volume of the mobile phase flowing from the injection time until the corresponding retention time of a molecular species and are related by,

$$V_R = Ut_R$$

The retention volume related to the dead time is known as dead volume V_0.

Migration rate

The migration rate can be defined as the velocity at which the species moves through the column. And the migration rate (U_R) is inversely proportional to the retention times. If only a fraction of molecules that are present in the mobile phase are moving. The value of migration rate is then given by,

$$U_R = u \times V_{mo}/(V_{mo} + V_{st})$$

Capacity factor

Capacity factor (k) is the ratio of reduced retention time and the dead time,

$$K = (t_R - t_0)t_0 = (V_R - V_0)/V_0$$

Equilibrium constant and phase ratio,

In the separation, the molecules running through the column can also be considered as being in a continuous equilibrium between the mobile phase and the stationary phase. This equilibrium could be governed by an equilibrium constant K, defined as,

$$K = C_{st}/C_{mo}$$

in which C_{mo} is the molar concentration of the molecules in the mobile phase, and C_{st} is the molar concentration of the molecules in the stationary phase. The equilibrium constant K can also be written as,

$$K = k(V_0/V_{st})$$

Advantage of HPLC

The most important aspect of HPLC is the high separation capacity which enables the batch analysis of multiple components. Even if the sample consists of a mixture, HPLC will allows the target components to be separated, detected, and quantified. Also, under appropriate condition, it is possible to attain a high level of reproducibility with a coefficient of variation not exceeding 1%. Also, it has a high sensitivity while a low sample consumption. HPLC has one advantage over GC column that analysis is possible for any sample can be stably dissolved in the eluent and need not to be vaporized. Thus, HPLC is used much more frequently in the field of biochemistry and pharmaceutical GC.

Bibliography

M. Serban and V. David, *Essentials in Modern HPLC Separation*, Elsevier, Waltham (2013).

S. Fanali, P. Hadded, C. Poole, P. Schoenmakers, and D. Lloyd, *Liquid Chromatography Fundamentals and Instrumentation*, Elsevier, Waltham (2013).

L. R. Snyder and J. J. Kirkland, *Introduction for Modern Liquid Chromatography*, Wily, New Jersey (2010).

R. P. W. Scott, *Liquid Chromatography Detectors*, Elsevier, New York (1986).

P. W. Scott, *Liquid Chromatography for the Analyst*, Marcel Dekker, New York (1994).

G. Schwedt, High-performance liquid chromatography in inorganic analysis. *Chromatographia*, 1979, **12**, 613.

Chapter 3: Supercritical Fluid Chromatography and Supercritical Fluid Extraction

Matthew Makansi, Mustafa Salih Hizir and Andrew R. Barron

Introduction

A popular and powerful tool in the chemical world, chromatography separates mixtures based on chemical properties even some than were previously thought inseparable. It combines a multitude of pieces, concepts, and chemicals to form an instrument suited to specific separation. However, the discovery of supercritical fluids led to novel analytical applications in the fields of chromatography and extraction known as supercritical fluid chromatography (SFC) and supercritical fluid extraction (SFE).

History

Supercritical fluid chromatography (SFC) begins its history in 1962 under the name high pressure gas chromatography and was quickly overshadowed by the development of high performance liquid chromatography (HPLC) and the already developed gas chromatography (GC). SFC was not a popular method of chromatography until the late 1980s, when more scientific publications began exemplifying its uses and techniques.

SFC was first reported by Ernst Klesper et al., where they succeeded in separating thermally labile porphyrin mixtures (Figure 3.1) on polyethylene glycol stationary phase with two mobile phase units: dichlorodifluoromethane (CCl_2F_2) and chlorodifluoromethane ($CHCl_2F$). Their results proved that supercritical fluids' low viscosity but high diffusivity functions well as a mobile phase.

After Klesper's paper detailing his separation procedure, subsequent scientists aimed to find the perfect mobile phase and the possible uses for SFC. Using gases such as He, N_2, CO_2, and NH_3, they examined purines, nucleotides, steroids, sugars, terpenes, amino acids, proteins, and many more substances for their retention behavior. They discovered that CO_2 was an ideal supercritical fluid due to its low critical temperature of 31 °C and relatively low critical pressure of 72.8 atm. Extra advantages of CO_2 included it being cheap, non-flammable, and non-toxic. CO_2 is now the standard mobile phase for SFC.

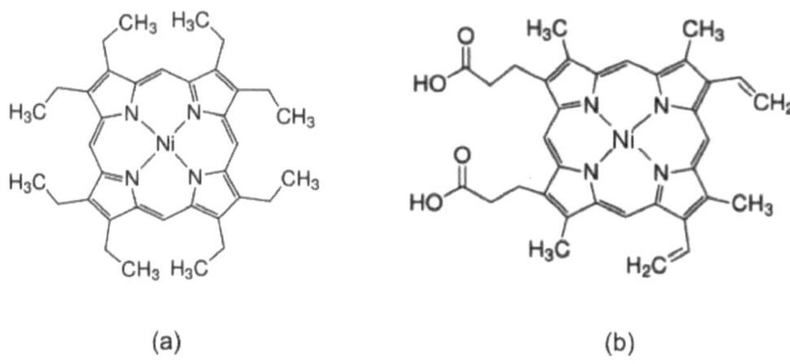

Figure 3.1: Thermally labile porphyrins (a) nickel etioporphyrin II and (b) nickel mesoporphyrin IX.

In the development of SFC over the years, the technique underwent multiple trial-and-error phases. Open tubular capillary column SFC had the advantage of independently and cooperatively changing all three parameters (pressure, temperature, and modifier content) to a certain extent. Like any chromatography method, however, it had its drawbacks. Changing the pressure, the most important parameter, often required changing the flow velocity due to the constant diameter of the capillaries. Additionally, CO_2, the ideal mobile phase, is non-polar, and its polarity could not be altered easily or with a gradient.

Over the years, many uses were discovered for SFC. It was identified as a useful tool in the separation of chiral compounds, drugs, natural products, and organometallics (see below for more detail). Most SFCs currently are involved a silica (or silica + modifier) packed column with a CO_2 (or CO_2 + modifier) mobile phase. Mass spectrometry is the most common tool used to analyze the separated samples.

Supercritical fluids

What is a supercritical fluid?

A supercritical fluid is the phase of a material at its critical temperature and critical pressure. Critical temperature (T_c) is the temperature above which it cannot be liquefied by pressure alone, and critical pressure (P_c) is the minimum amount of pressure to liquefy a gas at its critical temperature. Supercritical fluids combine useful properties of gas and liquid phases, as it can behave like both a gas and a liquid in terms of different aspects. Supercritical fluids are gas-like in the ways of expanding to fill a given volume, and the motions of the particles are close to that of a gas. On the side of liquid properties, supercritical fluids have densities near that of liquids and thus dissolve and interact with other particles, as you would expect of a liquid.

The formation of a supercritical fluid is the result of a dynamic equilibrium. When a material is heated to its specific critical temperature in a closed system, at constant pressure, a dynamic equilibrium is generated. This equilibrium includes the same number of molecules coming out of liquid phase to gas phase by gaining energy and going into liquid phase from gas phase by losing energy. At this particular point, the phase curve between liquid and gas phases disappears and supercritical material appears.

In order to understand the definition of SF better, a simple phase diagram can be used. Figure 3.2 displays an ideal phase diagram. For a pure material, a phase diagram shows the fields where the material is in the form of solid, liquid, and gas in terms of different temperature and pressure values. Curves, where two phases (solid-gas, solid-liquid and liquid-gas) exist together, defines the boundaries of the phase regions. These curves, for example, include sublimation for solid-gas boundary, melting for solid-liquid boundary, and vaporization for liquid-gas boundary. Other than these binary existence curves, there is a point where all three phases are present together in equilibrium; the triple point (TP). This is the temperature and pressure at which all three states can exist in a dynamic equilibrium.

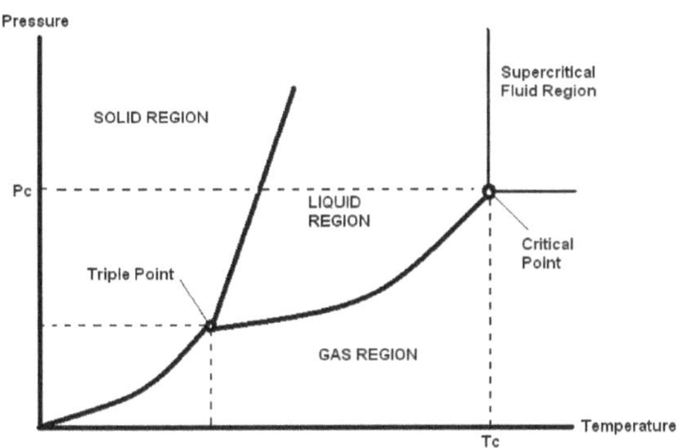

Figure 3.2: A generic phase diagram (with relevant points labeled).

There is another characteristic point in the phase diagram, the critical point (CP). This point is obtained at critical temperature (T_c) and critical pressure

(P_c). After the CP, no matter how much pressure or temperature is increased, the material cannot transform from gas to liquid or from liquid to gas phase. This form is the supercritical fluid form. Increasing temperature cannot result in turning to gas and increasing pressure cannot result in turning to liquid at this point. In the phase diagram, the field above T_c and P_c values is defined as the supercritical region.

In theory, the supercritical region can be reached in two ways:
- Increasing the pressure above the P_c value of the material while keeping the temperature stable and then increasing the temperature above T_c value at a stable pressure value.
- Increasing the temperature first above T_c value and then increasing the pressure above P_c value. The critical point is characteristic for each material, resulting from the characteristic T_c and P_c values for each substance.

Physical properties of supercritical fluids

The characteristic properties of a supercritical fluid are density, diffusivity and viscosity. Supercritical values for these features take place between liquids and gases. Table 3.1 demonstrates numerical values of properties for gas, supercritical fluid and liquid.

	Gas	Supercritical fluid	Liquid
Density (g/cm^3)	0.6×10^{-3} - 2.0×10^{-3}	0.2 - 0.5	0.6 - 2.0
Diffusivity (cm^2/s)	0.1 - 0.4	10^{-3} - 10^{-4}	0.2×10^{-5} - 2.0×10^{-2}
Viscosity (cm/s)	1×10^{-4} - 3×10^{-4}	1×10^{-4} - 3×10^{-4}	0.2×10^{-2} - 3.0×10^{-2}

Table 3.1: Supercritical fluid properties compared to liquids and gases.

Density

Density characteristic of a supercritical fluid is between that of a gas and a liquid, but closer to that of a liquid. In the supercritical region, density of a supercritical fluid increases with increased pressure (at constant temperature). When pressure is constant, density of the material decreases with increasing temperature. The dissolving effect of a supercritical fluid is dependent on its density value. Supercritical fluids are also better carriers than gases thanks to their higher density. Therefore, density is an essential parameter for analytical techniques using supercritical fluids as solvents.

Diffusivity

Diffusivity of a supercritical fluid can be 100 times that of a liquid and $^1/_{1,000}$ – $^1/_{10,000}$ times less than that of a gas. Because supercritical fluids have more diffusivity than a liquid, it stands to reason a solute can show better diffusivity in a supercritical fluid than in a liquid. Diffusivity is parallel with temperature and contrary with pressure. Increasing pressure affects supercritical fluid molecules to become closer to each other and decreases diffusivity in the material. The greater diffusivity gives supercritical fluids the chance to be faster carriers for analytical applications. Hence, supercritical fluids play an important role for chromatography and extraction methods.

Viscosity

Viscosity for a supercritical fluid is almost the same as a gas, being approximately $^1/_{10}{}^{th}$ of that of a liquid. Thus, supercritical fluids are less resistant than liquids towards components flowing through. The viscosity of supercritical fluids is also distinguished from that of liquids in that temperature has a little effect on liquid viscosity, where it can dramatically influence supercritical fluid viscosity.

Supercritical Properties of CO_2

Because of its widespread use in SFC, it's important to discuss what makes CO_2 an ideal supercritical fluid. One of the biggest limitations to most mobile phases in SFC is getting them to reach the critical point. This means extremely high temperatures and pressures, which is not easily attainable. The best gases for this are ones that can achieve a critical point at relatively low temperatures and pressures. As seen from Figure 3.3, CO_2 has a critical temperature of approximately 31 °C and a critical pressure of around 73 atm. These are both relatively low numbers and are thus ideal for SFC.

Of course, with every upside there exists a downside. In this case, CO_2 lacks polarity, which makes it difficult to use its mobile phase properties to elute polar samples. This is readily fixed with a modifier, which will be discussed later.

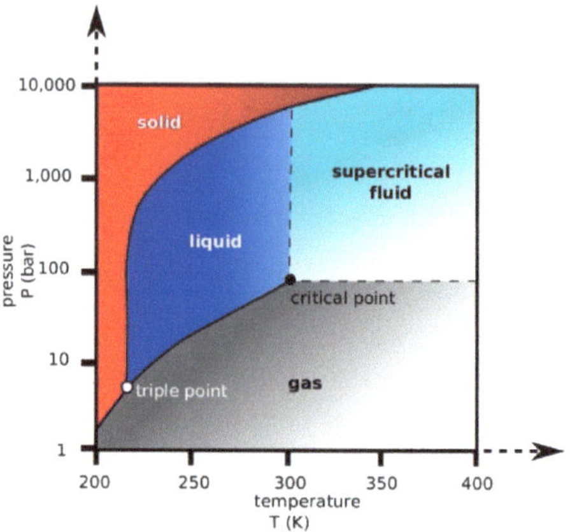

Figure 3.3: Phase diagram of CO₂.

Supercritical fluid chromatography (SFC)

Just like supercritical fluids combine the benefits of liquids and gases, SFC bring the advantages and strong aspects of HPLC and GC together. SFC can be more advantageous than HPLC and GC when compounds which decompose at high temperatures with GC and do not have functional groups to be detected by HPLC detection systems are analyzed.

There are three major qualities for column chromatographies:
- selectivity,
- efficiency,
- sensitivity.

Generally, HPLC has better selectivity that SFC flowing to changeable mobile phases (especially during a particular experimental run) and a wide range of stationary phases. Although SFC does not have the selectivity of HPLC, it has good quality in terms of sensitivity and efficiency. SFC enables change of some properties during the chromatographic process. This tuning ability allows the optimization of the analysis. Also, SFC has a broader range of detectors than HPLC. SFC surpasses GC for the analysis of easily decomposable substances; these materials can be used with SFC due to its ability to work with lower temperatures than GC.

Instrumentation for SFC

As it can be seen in Figure 3.4, SFC has a similar setup to an HPLC instrument. They use similar stationary phases with similar column types, however, there are some differences. Temperature is critical for supercritical fluids, so there should be a heat control tool in the system similar to that of GC. Also, there should be a pressure control mechanism, a restrictor, because pressure is another essential parameter in order for supercritical fluid materials to be kept at the required level. A microprocessor mechanism is placed in the instrument for SFC. This unit collects data for pressure, oven temperature, and detector performance to control the related pieces of the instrument.

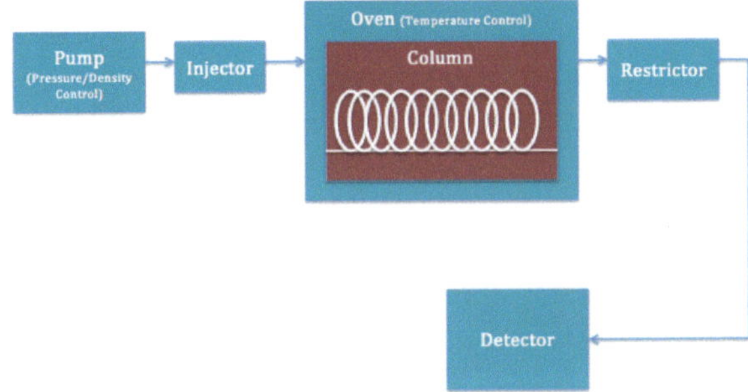

Figure 3.4: Scheme of a supercritical fluid chromatography instrument.

Pump

The existence of a supercritical fluid, as discussed previously, depends on high temperatures and high pressures. The pump is responsible for delivering the high pressures. By pressurizing the gas (or liquid), it can cause the substance to become dense enough to exhibit signs of the desired supercritical fluid. Because pressure couples with heat to create the supercritical fluid, the two are usually very close together on the instrument.

Injector

Injectors act as the main site for the insertion of samples. There are many different kinds of injectors that depend on a multitude of factors. For packed columns, the sample must be small, and the exact amount depends on the column diameter. For open tubular columns, larger volumes can be used. In both cases, there are specific injectors that are used depending on how the sample needs to be placed in the instrument. A loop injector is used mainly for preliminary testing. The sample is fed into a chamber that is then flushed with the supercritical fluid and pushed down the column. It uses a low-pressure pump before proceeding with the full elution at higher pressures. An inline injector allows for easy control of sample volume. A high-pressure pump forces the (specifically measured) sample into a stream of eluent, which proceeds to carry the sample through the column. This method allows for specific dilutions and greater flexibility. For samples requiring no dilution or immediate interaction with the eluent, an in-column injector is useful. This allows the sample to be transferred directly into the packed column and the mobile phase to then pass through the column.

Oven

The oven, as referenced before, exists to heat the mobile phase to its desired temperature. In the case of SFC, the desired temperature is always the critical temperature of the supercritical fluid. These ovens are precisely controlled and standard across SFC, HPLC, and GC.

Restrictor

As suggested by its name, the restrictor aims to restrict the flow through the columns. In altering the flow, the speed at which the sample elutes, the resolution at which it elutes, and the properties of the supercritical fluid can be altered, thus allowing for each SFC column to be tailored to the sample in question.

Detector

So far, there has been one largely overlooked component of the SFC machine: the detector. Technically not a part of the chromatographic separation process, the detector still plays an important role: identifying the components of the solution. While the SFC aims to separate components with good resolution (high purity, no other components mixed in), the detector aims to de ne what each of these components is made of.

The two detectors most often found on SFC instruments are either flame ionization detectors (FID) or mass spectrometers (MS):

FIDs operate through ionizing the sample in a hydrogen-powered flame. By doing so, they produce charged particles, which hit electrodes, and the particles are subsequently quantified and identified.

MS operates through creating an ionized spray of the sample, and then separating the ions based on a mass/charge ratio. The mass/charge ratio is plotted against ion abundance and creates a fingerprint for the chemical identified. This chemical fingerprint is then matched against a database to isolate which compound it was. This can be done for each unique elution, rendering the SFC even more useful than if it were standing alone.

Columns and stationary phase

There are two main types of columns used with SFC:
- open tubular,
- packed.

SFC columns are similar to HPLC columns in terms of coating materials. Open-tubular columns and packed columns are the two most common types used in SFC. Open-tubular ones are preferred, and they have similarities to HPLC fused-silica columns, in that they are coated with a cross-linked silica material (powdered quartz, SiO_2) for a stationary phase. This type of column contains an internal coating of a cross-linked siloxane material as a stationary phase. The thickness of the coating can be 0.05-1.0 μm. Column lengths range, but usually fall between 10 - 20 m. Figure 3.5 demonstrates the differences in the packing of the two columns.

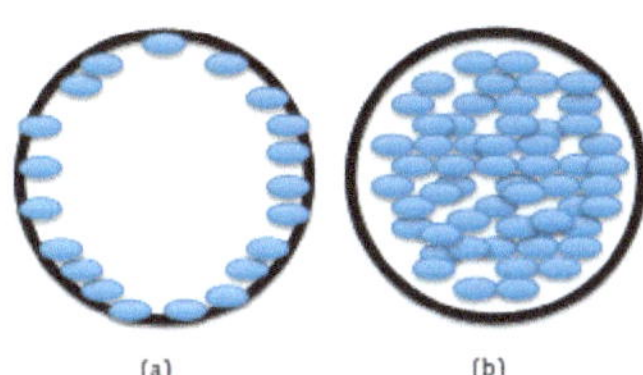

Figure 3.5: Schematic visualization of the difference between (a) open tubular and (b) packed column.

Mobile phase

The mobile phase (the supercritical fluid) pushes the sample through the column and elutes separate, pure, samples. There is a wide variety of materials used as mobile phase in SFC. The mobile phase can be selected from the solvent groups of inorganic solvents, hydrocarbons, alcohols, ethers, halides; or can be acetone, acetonitrile, pyridine, etc.

Solvent	Critical temperature (°C)	Critical pressure (bar)
Carbon dioxide (CO_2)	31.1	72
Nitrous oxide (N_2O)	36.5	70.6
Ammonia (NH_3)	132.5	109.8
Ethane (C_2H_6)	32.3	47.6
n-Butane (C_4H_{10})	152	70.6
Diethyl ether (Et_2O)	193.6	63.8
Tetrahydrofuran (THF, C_4H_8O)	267	50.5
Dichlorodifluoromethane (CCl_2F_2)	111.7	109.8

Table 3.2: Properties of some solvents as mobile phase at the critical point.

The most common supercritical fluid, which is used in SFC is CO_2 because its critical temperature and pressure are easy to reach. Additionally, CO_2 is low-cost, easy to obtain, inert towards UV, non-poisonous and a good solvent for non-polar molecules. Other than CO_2, ethane, n-butane, N_2O, dichlorodifluoromethane, diethyl ether, ammonia, and tetrahydrofuran can be used. Table 3.2 shows select solvents and their T_c and P_c values.

Modifiers

Modifiers are added to the mobile phase to play with its properties. As mentioned previously, CO_2 supercritical fluid lacks polarity. In order to add polarity to the fluid (without causing reactivity), a polar modifier will often be added. Modifiers usually raise the critical pressure and temperature of the mobile phase a little, but in return add polarity to the phase and result in a fully resolved sample. Unfortunately, with too much modifier, higher temperatures and pressures are needed and reactivity increases (which is dangerous and bad for the operator). Modifiers, such as ethanol or methanol, are used in small amounts as needed for the mobile phase in order to create a more polar fluid.

Sample

Generally speaking, samples need little preparation. The only major requirement is that it dissolves in a solvent less polar than methanol: it must have a dielectric constant lower than 33, since CO_2 has a low polarity and cannot easily elute polar samples. To combat this, modifiers are added to the mobile phase.

Detectors

One of the biggest advantages of SFC over HPLC is the range of detectors. Flame ionization detector (FID), which is normally present in GC setup, can also be applied to SFC. Such a detector can contribute to the quality of analyses of SFC since FID is a highly sensitive detector. SFC can also be coupled with a mass spectrometer, an UV-visible spectrometer, or an IR spectrometer more easily than can be done with an HPLC. Some other detectors which are used with HPLC can be attached to SFC such as fluorescence emission spectrometer or thermionic detectors.

Advantages of supercritical fluid chromatography

The physical properties of supercritical fluids between liquids and gases enables the SFC technique to combine with the best aspects of HPLC and GC, as lower viscosity of supercritical fluids makes SFC a faster method than HPLC. Lower viscosity leads to high flow speed for the mobile phase.

Thanks to the critical pressure of supercritical fluids, some fragile materials that are sensitive to high temperature can be analyzed through SFC. These materials can be compounds which decompose at high temperatures or materials which have low vapor pressure/volatility such as polymers and large biological molecules. High pressure conditions provide a chance to work with lower temperature than normally needed. Hence, the temperature-sensitive components can be analyzed via SFC. In addition, the diffusion of the components flowing through a supercritical fluid is higher than observed in HPLC due to the higher diffusivity of supercritical fluids over traditional liquids mobile phases. This results in better distribution into the mobile phase and better separation.

Clearly, SFC possesses some extraordinary potential as far as chromatography techniques go. It has some incredible capabilities that allow efficient and accurate resolution of mixtures. Below is a summary of its advantages and disadvantages stacked against other conventional (competing) chromatography methods.

Advantages over HPLC

- Because supercritical fluids have low viscosities the analysis is faster, there is a much lower pressure drop across the column, and open tubular columns can be used.

- Shorter column lengths are needed (10 - 20 m for SFC versus 15 - 60 m for HPLC) due to the high diffusivity of the supercritical fluid. More interactions can occur in a shorter span of time/distance.
- Resolving power is much greater (5×) than HPLC due to the high diffusivity of the supercritical fluid. More interactions result in better separation of the components in a shorter amount of time.

Advantages over GC

- Able to analyze many solutes with no derivatization since there is no need to convert most polar groups into nonpolar ones.
- Can analyze thermally labile compounds more easily with high resolution since it can provide faster analysis at lower temperatures.
- Can analyze solutes with high molecular weight due to their greater solubilizing power.

General disadvantages

- Cannot analyze extremely polar solutes due to relatively nonpolar mobile phase, CO_2.

Applications of SFC

The applications of SFC range from food to environmental to pharmaceutical industries. In this manner, pesticides, herbicides, polymers, explosives and fossil fuels are all classes of compounds that can be analyzed. SFC can be used to analyze a wide variety of drug compounds such as antibiotics, prostaglandins, steroids, taxol, vitamins, barbiturates, non-steroidal anti-inflammatory agents, etc. Chiral separations can be performed for many pharmaceutical compounds. SFC is dominantly used for non-polar compounds because of the low efficiency of carbon dioxide, which is the most common supercritical fluid mobile phase, for dissolving polar solutes. SFC is used in the petroleum industry for the determination of total aromatic content analysis as well as other hydrocarbon separations.

While the use of SFC has been mainly organic-oriented, there are still a few ways that inorganic compound mixtures are separated using the method. The two main ones, separation of chiral compounds (mainly metal-ligand complexes) and organometallics are discussed here.

Chiral compounds

For chiral molecules, the procedures and choice of column in SFC are very similar to those used in HPLC. Packed with cellulose type chiral stationary phase (or some other chiral stationary phase), the sample flows through the chiral compound and only molecules with a matching chirality will stick to the column. By running a pure CO_2 supercritical fluid mobile phase, the non-sticking enantiomer will elute first, followed eventually (but slowly) with the other one.

In the field of inorganic chemistry, a racemic mixture of Co(acac)₃, both isomers shown in Figure 3.6, has been resolved using a cellulose-based chiral stationary phase. The SFC method was one of the best and most efficient instruments in analyzing the chiral compound. While SFC easily separates coordinate covalent compounds, it is not necessary to use such an extensive instrument to separate mixtures of it since there are many simpler techniques.

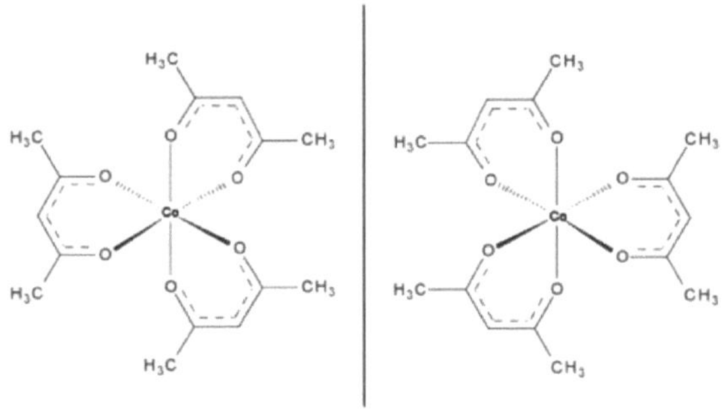

Figure 3.6: The two isomers of Co(acac)₃ in a racemic mixture which can be resolved by SFC.

Organometallics

Many d-block organometallics are highly reactive and easily decompose in air. SFC offers a way to chromatograph mixtures of large, unusual organometallic compounds. Large cobalt and rhodium based organometallic compound mixtures have been separated using SFC (Figure 3.7) without exposing the compounds to air.

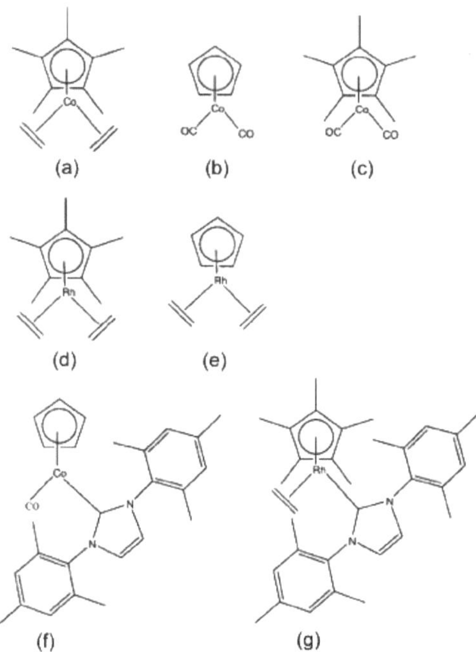

Figure 3.7: Examples of cobalt and rhodium based organometallic compound mixtures separated by SFC. Adapted from I. Bruheim, E. Fooladi, E. Lundanes, and T. Greibrokk, Purity testing of air-sensitive organometallic compounds by capillary supercritical fluid chromatography. *J. Microcolumn Sep.*, 2001, 13, 156.

By using a stationary phase of siloxanes, oxygen-linked silicon particles with different substituents attached, the organometallics were resolved based on size and charge. Thanks to the non-polar, highly diffusive, and high viscosity properties of a 100% CO_2 supercritical fluid, the mixture was resolved and analyzed with a flame ionization detector. It was determined that the method was sensitive enough to detect impurities of 1%. Because the efficiency of SFC is so impressive, the potential for it in the organometallic field is huge. Identifying impurities down to 1% shows promise for not only preliminary data in experiments, but quality control as well.

Supercritical fluid extraction (SFE)

The unique physical properties of supercritical fluids, having values for density, diffusivity and viscosity values between liquids and gases, enables supercritical fluid extraction to be used for the extraction processes which cannot be done by liquids due to their high density and low diffusivity and by gases due to their inadequate density in order to extract and carry the components out.

Complicated mixtures containing many components should be subject to an extraction process before they are separated via chromatography. An ideal extraction procedure should be fast, simple, and inexpensive. In addition, sample loss or decomposition should not be experienced at the end of the extraction. Following extraction, there should be a quantitative collection of each component. Ideally, the amount of unwanted materials coming from the extraction should be kept to a minimum and be easily disposable; the waste should not be harmful for environment. Unfortunately, traditional extraction methods often do not meet these requirements. In this regard, SFE has several advantages in comparison with traditional techniques.

The extraction speed is dependent on the viscosity and diffusivity of the mobile phase. With a low viscosity and high diffusivity, the component which is to be extracted can pass through the mobile phase easily. The higher diffusivity and lower viscosity of supercritical fluids, as compared to regular extraction liquids, help the components to be extracted faster than other techniques. Thus, an extraction process can take just 10-60 minutes with SFE, while it would take hours or even days with classical methods.

The dissolving efficiency of a supercritical fluid can be altered by temperature and pressure. In contrast, liquids are not affected by temperature and pressure changes as much. Therefore, SFE has the potential to be optimized to provide a better dissolving capacity.

In classical methods, heating is required to get rid of the extraction liquid. However, this step causes the temperature-sensitive materials to decompose. For SFE, when the critical pressure is removed, a supercritical fluid transforms to gas phase. Because supercritical fluid solvents are chemically inert, harmless and inexpensive; they can be released to atmosphere without leaving any waste. Through this, extracted components can be obtained much more easily and sample loss is minimized.

Instrumentation for SFE

The necessary apparatus for SFE is simple. Figure 3.8 depicts the basic elements of the SFE instrument, which is composed of a reservoir of supercritical fluid, a pressure tuning injection unit, two pumps (to take the components in the mobile phase in and to send them out of the extraction cell), and a collection chamber.

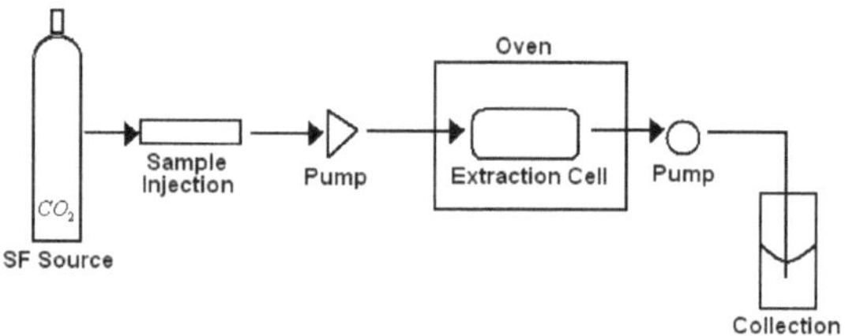

Figure 3.8: Scheme of an idealized supercritical fluid extraction instrument.

There are two principle modes to run the instrument:
- Static extraction.
- Dynamic extraction.

In dynamic extraction, the second pump sending the materials out to the collection chamber is always open during the extraction process. Thus, the mobile phase reaches the extraction cell and extracts components in order to take them out consistently.

In the static extraction experiment, there are two distinct steps in the process:
- The mobile phase fills the extraction cell and interacts with the sample.
- The second pump is opened, and the extracted substances are taken out at once.

In order to choose the mobile phase for SFE, parameters taken into consideration include the polarity and solubility of the samples in the mobile phase. Carbon dioxide is the most common mobile phase for SFE. It has a capability to dissolve non-polar materials like alkanes. For semi-polar compounds (such as polycyclic aromatic hydrocarbons, aldehydes, esters, alcohols, etc.) CO_2 can be used as a single component mobile phase. However, for compounds which have polar characteristic, supercritical carbon dioxide must be modified by addition of polar solvents like methanol (CH_3OH). These extra solvents can be introduced into the system through a separate injection pump.

Extraction modes

There are two modes in terms of collecting and detecting the components:
- Off-line extraction.
- On-line extraction.

Off-line extraction is done by taking the mobile phase out with the extracted components and directing them towards the collection chamber. At this point, supercritical fluid phase is evaporated and released to atmosphere and the components are captured in a solution or a convenient adsorption surface. Then the extracted fragments are processed and prepared for a separation method. This extra manipulation step between extractor and chromatography instrument can cause errors. The on-line method is more sensitive because it directly transfers all extracted materials to a separation unit, mostly a chromatography instrument, without taking them out of the mobile phase. In this extraction/detection type, there is no extra sample preparation after extraction for separation process. This minimizes the errors coming from manipulation steps. Additionally, sample loss does not occur and sensitivity increases.

Applications of SFE

SFE can be applied to a broad range of materials such as polymers, oils and lipids, carbonhydrates, pesticides, organic pollutants, volatile toxins, polyaromatic hydrocarbons, biomolecules, foods, flavors, pharmaceutical metabolites, explosives, and organometallics, etc. Common industrial applications include the pharmaceutical and biochemical industry, the polymer industry, industrial synthesis and extraction, natural product chemistry, and the food industry.

Examples of materials analyzed in environmental applications: oils and fats, pesticides, alkanes, organic pollutants, volatile toxins, herbicides, nicotin, phenanthrene, fatty acids, aromatic surfactants in samples from clay to petroleum waste, from soil to river sediments. In food analyses: caffeine, peroxides, oils, acids, and cholesterol, etc., are extracted from samples such as coffee, olive oil, lemon, cereals, wheat, potatoes and dog feed. Through industrial applications, the extracted materials vary from additives to different oligomers, and from petroleum fractions to stabilizers. Samples analyzed are plastics, PVC, paper, wood etc. Drug metabolites, enzymes, steroids are extracted from plasma, urine, serum or animal tissues in biochemical applications.

Bibliography

M. Caude and D. Thiebaut, *Practical Supercritical Fluid Chromatography and Extraction*, Harwood Academic Publishers, Switzerland (1999).

T. L. Chester and J. D. Pinkston, Supercritical-fluid chromatography. *Anal. Chem.*, 1990, **62**, 394.

D. R. Gere, R. Board, and D. McManigill, Supercritical fluid chromatography with small particle diameter packed columns. *Anal. Chem.*, 1982, **54**, 736.

E. Klesper, A. H. Corwin, and D. A. Turner, High pressure gas chromatorgraphy above critical temperatures. *Org. Chem.*, 1962, **27**, 700.

M. D. Luque de Castro, M. Valcarcel, and M. T. Tena, *Analytical Supercritical Fluid Extraction*, Springer-Verlag, Berlin (1994).

M.N. Meyers, J. Giddings, High column efficiency in gas liquid chromatography at inlet pressures to 2500 psi. *Anal. Chem.*, 1965, **37**, 1453.

K. Miyazawa, T. Ishiguro, and H. Oda, *FPC production*. Daicel Chemical Industries, LTD. Myoko, Japan (2007).

M. D. Palmieri, An introduction to supercritical fluid chromatography: part 2. applications and future trends. *J. Chem. Educ.*, 1989, **66**, A141.

M. Saito, History of supercritical fluid chromatography: Instrumental development. *J. Biosci. Bioeng.*, 2013, **115**, 590.

D. A. Skoog and J. J. Leary, *Principles of Instrumental Analysis*, Saunders College Publishing, Philadelphia (1992).

R. M. Smith, *Supercritical Fluid Chromatography*, Royal Society of Chemistry, Cambridge (1988).

L. T. Taylor, *Supercritical Fluid Extraction*, Wiley-Interscience Publication, New York, (1996).

L. Toribio, C. Alonso, M. J. del Nozal, J. L. Bernal, and J. J. Jimnez, Enantiomeric separation of chiral sulfoxides by supercritical fluid chromatography. *J. Sep. Sci.*, 2006, **29**, 1363.

B. W. Wenclawiak, Ernst Klesper, the "father of supercritical fluid chromatography". *Fresenius J. Anal. Chem.*, 1992, **344**, 425.

M. Yoshioka, S. Parvez, T. Miazaki, and H. Parvez, *Supercritical Fluid Chromatography and Micro-HPLC*, VNU Science Press, The Netherlands (1989).

Chapter 4: Ion Chromatography

Brett Virgin-Downey and Andrew R. Barron

Introduction

Ion Chromatography is a method of separating ions based on their distinct retention rates in a given solid phase packing material. Given different retention rates for two anions or two cations, the elution time of each ion will differ, allowing for detection and separation of one ion before the other. Detection methods are separated between electrochemical methods and spectroscopic methods. This guide will cover the principles of retention rates for anions and cations, as well as describing the various types of solid-state packing materials and eluents that can be used.

Principles of ion chromatography

Retention models in anion chromatography

The retention model for anionic chromatography can be split into two distinct models, one for describing eluents with a single anion, and the other for describing eluents with complexing agents present. Given an eluent anion or an analyte anion, two phases are observed, the stationary phase (denoted by S) and the mobile phase (denoted by M). As such, there is equilibrium between the two phases for both the eluent anions and the analyte anions that can be described by,

$$y * [A_M^{x-}] + x * [E_S^{y-}] \Leftrightarrow y * [A_S^{x-}] + x * [E_M^{y-}]$$

This yields an equilibrium constant as given in,

$$K_{A,E} = \frac{[A_S^{x-}]^y [E_M^{y-}]^x \gamma_{A_S^{x-}}^y \gamma_{E_M^{y-}}^x}{[A_M^{x-}]^y [E_S^{y-}]^x \gamma_{A_M^{x-}}^y \gamma_{E_S^{y-}}^x}$$

Given the activity of the two ions cannot be found in the stationary or mobile phases, the activity coefficients are set to 1. Two new quantities are then introduced. The first is the distribution coefficient, D_A, which is the ratio of analyte concentrations in the stationary phase to the mobile phase,

$$D_A = \frac{[A_S]}{[A_M]}$$

The second is the retention factor, k^1,

$$k_A^1 = D_A * \frac{V_S}{V_M}$$

which is the distribution coefficient multiplied by the ratio of volume between the two phases, Substituting the two quantities, the equilibrium constant can be written as:

$$K_{A,E} = \left(k_A^1 \frac{V_M}{V_S}\right)^y * \left(\frac{[E_M^{y-}]}{[E_S^{y-}]}\right)^x$$

Given there is usually a large difference in concentrations between the eluent and the analyte (with magnitudes of 10 greater eluent), this equation can be re-written under the assumption that all the solid phase packing material's functional groups are taken up by E^{y-}. As such, the stationary E^{y-} can be substituted with the exchange capacity divided by the charge of E^{y-},

$$K_{A,E} = \left(k_A^1 \frac{V_M}{V_S}\right)^y * \left(\frac{Q}{Y}\right)^{-x} [E_M^{y-}]^x$$

Solving for the retention factor,

$$z * [A_M^{x-}] + x * [B_S^{z-}] \Leftrightarrow z * [A_S^{x-}] + x * [B_M^{z-}]$$

which shows the relationship between retention factor and parameters like eluent concentration and the exchange capacity, which allows parameters of the ion chromatography to be manipulated and the retention factors to be determined. This only works for a single analyte present, but a relationship for the selectivity between two analytes [A] and [B] can easily be determined.

First the equilibrium between the two analytes is determined as

$$K_{A,B} = \frac{[A_S^{x-}]^z [B_M^{z-}]^x}{[A_M^{x-}]^z [B_S^{z-}]^x}$$

The equilibrium constant (ignoring activity) can be written as:

$$\alpha_{A,B} = \frac{[A_S^{x-}][B_M^{z-}]}{[A_M^{x-}][B_S^{z-}]}$$

The selectivity can then be determined to be:

$$\alpha_{A,B} = \frac{[A_S^{x-}][B_M^{z-}]}{[A_M^{x-}][B_S^{z-}]}$$

can then be simplified into a logarithmic form as the following two equations:

$$\log \alpha_{A,B} = \frac{1}{z} \log K_{A,B} + \frac{x-z}{z} \log \frac{k_b^1 V_M}{V_S}$$

$$\log \alpha_{A,B} = \frac{1}{x} \log K_{A,B} + \frac{x-z}{z} \log \frac{k_A^1 V_M}{V_S}$$

When the two charges are the same, it can be seen that the selectivity is only a factor of the selectivity coefficients and the charges. When the two charges are different, it can be seen that the two retention factors are dependent upon each other.

In situations with a polyatomic eluent, three models are used to account for the multiple anions in the eluent. The first is the dominant equilibrium model, in which one anion is so dominant in concentration; the other eluent anions are ignored. The dominant equilibrium model works best for multivalence analytes. The second is the effective charge model, where an effective charge of the eluent anions is found, and a relationship is found with the effective charge. The effective charge models work best with monovalent analytes. The third is the multiple eluent species model, where

$$\log K_A^1 = C_3 - \left(\frac{X_1}{a} + \frac{X_2}{b} + \frac{X_3}{c}\right) - \log C_p$$

describes the retention factor. C_3 is a constant that includes the phase volume ratio between stationary, the equilibrium constant, and mobile and the exchange capacity. C_p is the total concentration of the eluent species. X_1, X_2, and X_3 correspond to the shares of a particular eluent anion in the retention of the analyte.

Retention models of cation chromatography

For eluents with a single cation and analytes that are alkaline earth metals, heavy metals or transition metals, a complexing agent is used to bind with the metal during chromatography. This introduces the quantity A(m) to the retention rate calculations, where A(m) is the ratio of free metal ion to the total concentration of metal. Following a similar derivation to the single anion case is found,

$$K_{A,E} = \left(\frac{k_A^1}{\alpha_M\,\phi}\right)^y * \left(\frac{Q}{Y}\right)^{-x} [E_M^{y+}]^x$$

Solving for the retention coefficient,

$$k_A^1 = \alpha_M\,\phi * K_{A,E}^{\frac{1}{y}} \left(\frac{Q}{Y}\right)^{\frac{x}{y}} ([E_M^{y+}])^{-\frac{x}{y}}$$

From this expression, the retention rate of the cation can be determined from eluent concentration and the ratio of free metal ions to the total concentration of the metal, which itself is depended on the equilibrium of the metal ion with the complexing agent.

Solid phase packing materials

The solid phase packing material used in the chromatography column is important to the exchange capacity of the anion or cation. There are many types of packing material, but all share a functional group that can bind either the anion or the cation complex. The functional group is mounted on a polymer surface or sphere, allowing large surface area for interaction.

Packing material for anion chromatography

The primary functional group used for anion chromatography is the ammonium group. Amine groups are mounted on the polymer surface, and the pH is lowered to produce ammonium groups. As such, the exchange capacity is depended on the pH of the eluent. To reduce the pH dependency, the protons on the ammonium are successively replaced with alkyl groups until the all the protons are replaced and the functional group is still positively charged, but pH independent. The two packing materials used in almost all anion chromatography are trimethyl amine and dimethylethanolamine derived functionality (Figure 4.1 and 4.2, respectively).

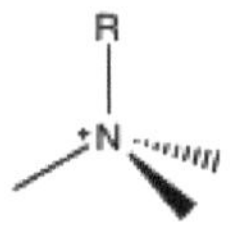

Figure 4.1: The trimethylammonium functionality mounted on a polymer used as a solid phase packing material.

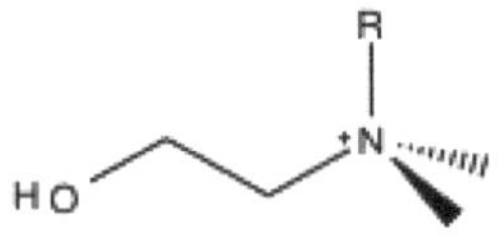

Figure 4.2: A dimethylethanol ammonium mounted on a polymer used as solid phase packing material.

Packing material for cation chromatography

Cation chromatography allows for the use of both organic polymer-based and silica gel-based packing material. In the silica gel-based packing material, the most common packing material is a polymer-coated silica gel. The silicate is coated in polymer, which is held together by cross-linking of the polymer. Polybutadiene maleic acid (Figure 4.3) is then used to create a weakly acidic material, allowing the analyte to di use through the polymer and exchange. Silica gel-based packing material is limited by the pH dependent solubility of the silica gel and the pH dependent linking of the silica gel and the functionalized polymer. However, silica gel-based packing material is suitable for separation of alkali metals and alkali earth metals.

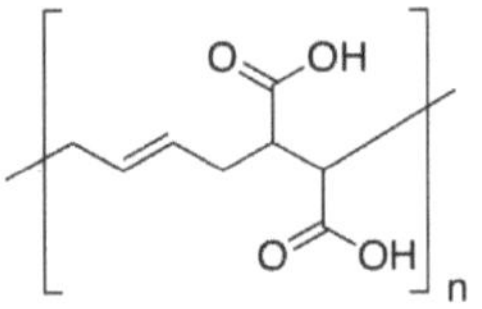

Figure 4.3: A polybutadiene maleic acid polymer used as a cation solid phase packing material.

Organic polymer-based packing material is not limited by pH like the silica gel materials are not suitable for separation of alkali metals and alkali earth metals. The most common functional group is the sulfonic acid group (Figure 4.4), attached with a spacer between the polymer and the sulfonic acid group.

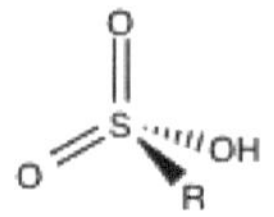

Figure 4.4: A sulfonic acid group used as a cation solid phase packing material functional group.

Detection methods

Spectroscopic detection methods

Photometric detection in the UV region of the spectrum is a common method of detection in ion chromatography. Photometric methods limit the eluent possibilities, as the analyte must have a unique absorbance wavelength to be detectable. Cations that do not have a unique absorbance wavelength, i.e., the eluent and other contaminants have similar UV visible spectra can be complexed to for UV visible compounds. This allows detection of the cation without interference from eluents.

Coupling the chromatography with various types of spectroscopy such as Mass spectroscopy or IR spectroscopy can be a useful method of detection. Inductively coupled plasma atomic emission spectroscopy is a commonly used method.

Direct conductivity methods

Direct conductivity methods take advantage of the change in conductivity that an analyte produces in the eluent, which can be modeled by,

$$\Delta K = \frac{(\Lambda_A - \Lambda_E) * C_S}{1000}$$

where equivalent conductivity is defined as,

$$\Lambda = \frac{L}{A * R} * \frac{1}{C}$$

With L being the distance between two electrodes of area A and R being the resistance the ion creates. C is the concentration of the ion. The conductivity can be plotted over time, and the peaks that appear represent different ions coming through the column as described by,

$$K_{peak} = (\Lambda_A - \Lambda_E) * C_A$$

The values of Equivalent conductivity of the analyte and of the eluent common ions can be found in Tables 4.1 and 4.2.

Cation	Λ^+ (S.cm^2/eq)	Anion	Λ^+ (S.cm^2/eq)
H^+	350	OH^-	198
Li^+	39	F^-	54
Na^+	50	Cl^-	76
K^+	74	Br^-	78
NH_4^+	73	I^-	77
$\frac{1}{2}Mg^{2+}$	53	NO_2^-	72
$\frac{1}{2}Ca^{2+}$	60	NO_3^-	71
$\frac{1}{2}Sr^{2+}$	59	HCO_3^-	45
$\frac{1}{2}Ba^{2+}$	64	$\frac{1}{2}CO_3^{2-}$	72
$\frac{1}{2}Zn^{2+}$	53	$H_2PO_4^-$	33
$\frac{1}{2}Hg^{2+}$	55	$\frac{1}{2}HPO_4^{2-}$	57
$\frac{1}{2}Cu^{2+}$	71	$\frac{1}{3}PO_4^{3-}$	69
$\frac{1}{2}Pb^{2+}$	53	$\frac{1}{2}SO_4^{2-}$	80
$\frac{1}{2}Co^{2+}$	70	CN^-	82
$1/_3Fe^{3+}$	33	SCN^-	66

Table 4.1: Equivalent conductivities of inorganic ions. Data from C. Eith, M. Kolb, A. Seubert, and K. H. Viehweger. *Practical Ion Chromatography: An Introduction*, Metrohm, Ltd., CH-9101 Herisau, Switzerland (2001).

Organic anion	Λ^+ (S.cm^2/eq)
Acetate	41
Propionate	38
½ Phthalate	36
Benzoate	32
Salicylate	30
½Oxalate	74

Table 4.2: Equivalent conductivities of organic ions (Figure 4.5). Data from C. Eith, M. Kolb, A. Seubert, and K. H. Viehweger. *Practical Ion Chromatography: An Introduction*, Metrohm, Ltd., CH-9101 Herisau, Switzerland (2001).

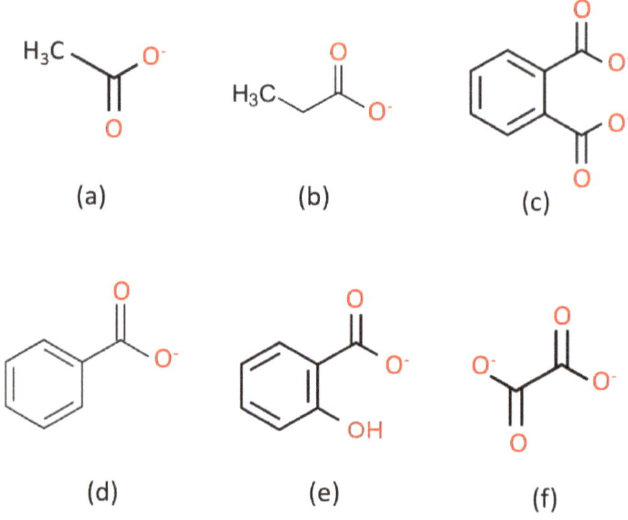

Figure 4.5: Structure of (a) acetate anion, CH₃CO₂⁻, (b) propionate anion, CH₃CH₂CO₂⁻, (c) phthalate dianion, C₆H₄(CO₂⁻)₂, (d) benzoate anion, C₆H₅CO₂⁻, (e) salicylate anion, C₆H₄(OH)CO₂⁻, and (f) oxalate dianion, (CO₂⁻)₂.

Eluents

The choice of eluent depends on many factors, namely, pH, buffer capacity, the concentration of the eluent, and the nature of the eluent's reaction with the column and the packing material.

Eluents in anion chromatography

In non-suppressed anion chromatography, where the eluent and analyte are not altered between the column and the detector, there is a wide range of eluents to be used. In the non-suppressed case, the only issue that could arise is if the eluent impaired the detection ability (absorbing in a similar place in a UV-spectra as the analyte for instance). As such, there are a number of commonly used eluents. Aromatic carboxylic acids are used in conductivity detection because of their low self-conductivity. Aliphatic carboxylic acids are used for UV-visible detection because they are UV transparent. Inorganic acids can only be used in photometric detection.

In suppressed anion chromatography, where the eluent and analyte are treated between the column and detection, fewer eluents can be used. The suppressor modifies the eluent and the analyte, reducing the self- conductivity of the eluent and possibly increasing the self-conductivity of the analyte. Only alkali hydroxides and carbonates, borates, hydrogen carbonates, and amino acids can be used as eluents.

Eluents in cation chromatography

The primary eluents used in cation chromatography of alkali metals and ammoniums are mineral acids such as HNO3. When the cation is multivalent, organic bases such as ethylenediamine (Figure 4.6) serve as the main eluents. If both alkali metals and alkali earth metals are present, hydrochloric acid or 2,3-diaminopropionic acid (Figure 4.7) is used in combination with a pH variation. If the chromatography is unsuppressed, the direct conductivity measurement of the analyte will show up as a negative peak due to the high conductivity of the H^+ in the eluent, but simple inversion of the data can be used to rectify this discrepancy.

Figure 4.6: Ethylenediamine, a commonly used eluent in cation chromatography.

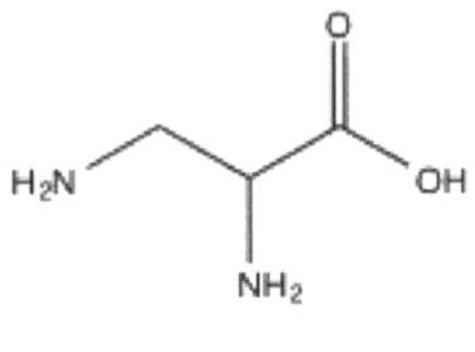

Figure 4.7: 2,3-diaminopropionic acid, a primary eluent for cation chromatography of alkali and alkali earth metal combinations.

If transition metals or H^+ are the analytes in question, complexing carboxylic acids are used to suppress the charge of the analyte and to create photometrically detectable complexes, forgoing the need for direct conductivity as the detection method.

Bibliography

I. Demkowska, Z. Polkowska, and J. Namiesnik, Application of ion chromatography for the determination of inorganic ions, especially thiocyanates in human saliva samples as biomarkers of environmental tobacco smoke exposure. *J. Chromatogr. B*, 2008, **875**, 419.

C. Eith, M. Kolb, A. Seubert. K.H. Viehweger. *Practical Ion Chromatography: An Introduction*. Metrohm, Ltd., Herisau, Switzerland (2001).

T. P. Moyer, Optimized isocratic conditions for analysis of catecholamines by high-performance reversed-phase paired-ion chromatography with amperometric detection. *J. Chromatogr. A*, 1978, **153**, 365.

Chapter 5: Capillary Electrophoresis (CE)

Mahnoor Rafik Madakia, Pavan M. V. Raja and Andrew R. Barron

Introduction

Capillary electrophoresis (CE) encompasses a family of electrokinetic separation techniques that uses an applied electric field to separate out analytes based on their charge and size. The basic principle is hinged upon that of electrophoresis, which is the motion of particles relative to a fluid(electrolyte) under the influence of an electric field.

The founding father of electrophoresis, Arne W. K. Tiselius (Figure 5.1), first used electrophoresis to separate proteins, and he went on to win a Nobel Prize in Chemistry in 1948 for his work on both electrophoresis and adsorption analysis. However, it was Stellan Hjerten (Figure 5.2), who worked under Arne W. K. Tiselius, who pioneered work in CE in 1967, although CE was not well recognized until 1980 when James W. Jorgenson (Figure 5.3) and Krynn D. Lukacs published a series of papers describing this new technique.

Figure 5.1: Swedish chemist Arne W. K. Tiselius (1902 - 1971) who was the founding father of electrophoresis.

Figure 5.2: Swedish chemist Stellan Hjerten (1928 -) who worked under Arne W. K. Tiselius that pioneered work in CE.

Figure 5.3: James W. Jorgensen (1952-).

Instrumentation overview

The main components of CE are shown in Figure 3.4. The electric circuit of the CE is the heart of the instrument.

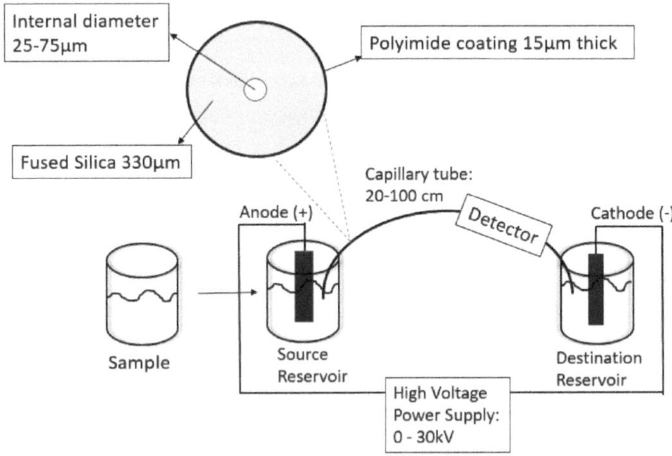

Figure 5.4: A schematic diagram of the components of a typical capillary electrophoresis setup and the capillary column.

Injection methods

The samples that are studied in CE are mainly liquid samples. A typical capillary column has an inner diameter of 50 μm and a length of 25 cm. Because the column can only contain a minimal amount of running buffer, only small sample volumes can be tested (nL to μL). The samples are introduced mainly by two injection methods: hydrodynamic and electrokinetic injection. The two methods are displayed in Table 5.1. A disadvantage of electrokinetic injection is that the composition of the injected sample may not be the same as the composition of the original sample. This is because the injection method is dependent on the electrophoretic and electroosmotic mobility of the species in the sample. However, both injection methods depend on the temperature and the viscosity of the solution. Hence, it is important to control both parameters when a reproducible volume of sample injections is desired. It is advisable to use internal standards instead of external standards when performing quantitative analysis on the samples as it is hard to control both the temperature and viscosity of the solution.

Injection method	Working Principle
Hydrody-namic injection	The sample vial is enclosed in a chamber with one end of capillary column immersed in it. Pressure is then applied to the chamber for a fixed period so that the sample can enter the capillary. After the sample, has been introduced, the capillary is withdrawn and then re-immersed into the source reservoir and separation takes place.
Electroki-netic injection	The sample is enclosed in a chamber with one end of capillary column immersed in it with an electrode present. The electric field is applied, and the samples enter the capillary. After the sample, has been introduced, the capillary is withdrawn and then re-immersed into the source reservoir and separation takes place.

Table 5.1: The working principle of the two injection methods used in CE.

Column

After the samples have been injected, the capillary column is used as the main medium to separate the components. The capillary column used in CE shares the same characteristics as the capillary column used in gas chromatography (GC); however, the most critical components of the CE column are:
- the inner diameter of the capillary,
- the total length of the capillary,
- the length of the column from the injector to the detector.

Solvent buffer

The solvent buffer carries the sample through the column. It is crucial to employ a good buffer as a successful CE experiment is hinged upon this. CE is based on the separation of charges in an electric field. Therefore, the buffer should either sustain the pre-existing charge on the analyte or enable the analyte to obtain a charge, and it is important to consider the pH of the buffer before using it.

Applied voltage (kV)

The applied voltage is important in the separation of the analytes as it drives the movement of the analyte. It is important that it is not too high as it may become a safety concern.

Detectors

Analytes that have been separated after the applying the voltage can be detected by many detection methods. The most common method is UV-visible absorbance. The detection takes place across the capillary with a small portion of the capillary acting as the detection cell. The on-tube detection cell is usually made optically transparent by scraping off the polyimide coating and coating it with another optically transparent material so that the capillary would not break easily. For species that do not have a chromophore, a chromophore can be added to the buffer solution. When the analyte passes by, there would be a decrease in signal. This decreased signal will correspond to the amount of analyte present. Other common detection techniques employable in CE are fluorescence and mass spectrometry (MS).

Theory

In CE, the sample is introduced into the capillary by the above-mentioned methods. A high voltage is then applied causing the ions of the sample to migrate towards the electrode in the destination reservoir, in this case, the cathode. Sample components migration and separation are determined by two factors, electrophoretic mobility and electroosmotic mobility.

Electrophoretic mobility

The electrophoretic mobility, μ_{ep}, is inherently dependent on the properties of the solute and the medium in which the solute is moving. Essentially, it is a constant value, that can be calculated as given by (3.31), where q is the solute's charge, η is the buffer viscosity and r is the solutes radius.

$$\mu_{ep} = q/6\pi\eta r$$

The electrophoretic velocity, v_{ep}, is dependent on the electrophoretic mobility and the applied electric field, E,

$$v_{ep} = \mu_{ep}E$$

Thus, when solutes have a larger charge to size ratio the electrophoretic mobility and velocity will increase. Cations and the anion would move in opposing directions corresponding to the sign of the electrophoretic mobility with is a result of their charge. Thus, neutral species that have no charge do not have an electrophoretic mobility.

Electroosmotic mobility

The second factor that controls the migration of the solute is the electroosmotic flow. With zero charge, it is expected that the neutral species should remain stationary. However, under normal conditions, the buffer solution moves towards the cathode as well. The cause of the electroosmotic flow is the electric double layer that develops at the silica solution interface.

At pH more than 3, the abundant silanol (-OH) groups present on the inner surface of the silica capillary, deprotonate to form negatively charged silanate ions (-SiO⁻). The cations present in the buffer solution will be attracted to the silanate ions and some of them will bind strongly to it forming a fixed layer. The formation of the fixed layer only partially neutralizes the negative charge on the capillary walls. Hence, more cations than anions will be present in the layer adjacent to the fixed layer, forming the di use layer. The combination of the fixed layer and di use layer is known as the double layer as shown in Figure 5.5. The cations present in the di use layer will migrate towards the cathode, as these cations are solvated the solution will also flow with it, producing the electroosmotic flow. The anions present in the di use layer are solvated and will move towards the anode. However, as there are more cations than anions the cations will push the anions together with it in the direction of the cathode. Hence, the electroosmotic flow moves in the direction of the cathode.

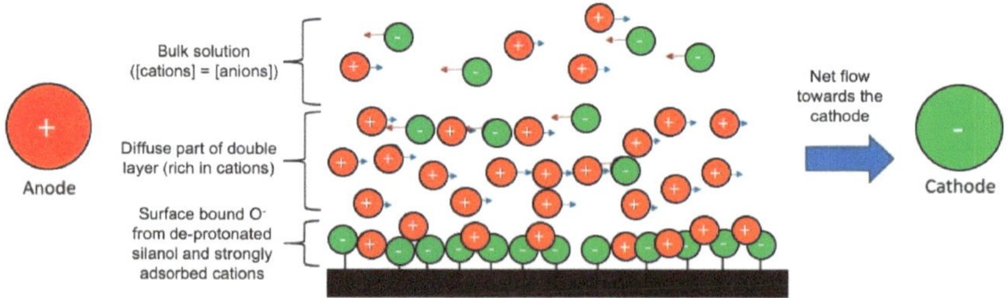

Figure 5.5: An illustration of the electric double layer and movement of the species in solution.

The electroosmotic mobility, μ_{eof}, is described by

$$\mu_{eof} = \xi\varepsilon/4\pi\eta$$

where ξ is the zeta potential, ε is the buffer dielectric constant and η is the buffer viscosity. The electroosmotic velocity, v_{eof}, is the rate at which the buffer moves through the capillary is given by

$$v_{eof} = \mu_{eof}E$$

Zeta potential

The zeta potential, ξ, also known as the electrokinetic potential is the electric potential at the interface of the double layer. Hence, in our case, it is the potential of the diffuse layer that is at a finite distance from the capillary wall. Zeta potential is mainly affected and directly proportional to two factors:

The thickness of the double layer. A higher concentration of cations possibly due to an increase in the buffer's ionic strength would lead to a decrease in the thickness of the double layer. As the thickness of the double layer decreases, the zeta potential would decrease that results in the decrease of the electroosmotic flow.

The charge on the capillary walls due to the greater density of the silanate ions ($-SiO^-$) corresponds to a larger zeta potential. The formation of silanate ions is pH dependent. Hence, at pH less than 2 there is a decrease in the zeta potential and the electroosmotic flow as the silanol exists in its protonated form. However, as the pH increases, there are more silanate ions formed causing an increase in zeta potential and hence, the electroosmotic flow.

Order of elution

Electroosmotic flow of the buffer is generally greater than the electrophoretic flow of the analytes. Hence, even the anions would move to the cathode as illustrated in Figure 5.6. Small, highly charged cations would be the first to elute before larger cations with lower charge. This is followed by the neutral species which elutes as one band in the middle. The larger anions with low charge elute next and lastly, the highly charged small anion would have the longest elution time. This is clearly portrayed in the electropherogram in Figure 5.7.

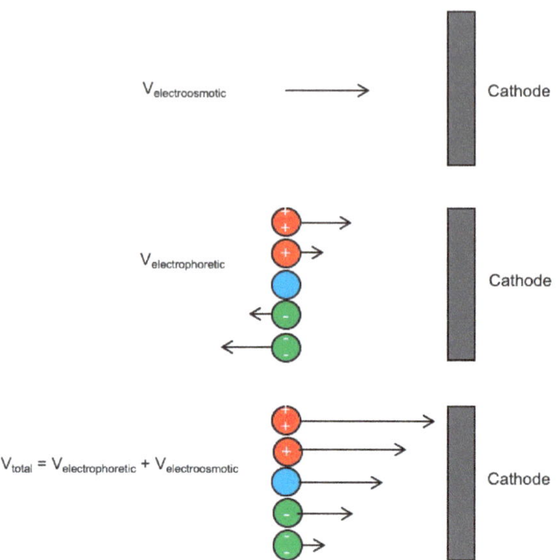

Figure 5.6: An illustration of the order of elution of the charged species as signified by the length of the arrows. Adapted from D. A. Skoog, D. M. West, F. J. Holler, and S. R. Crouch, *Fundamentals of Analytical Chemistry*, Brooks Cole (2013).

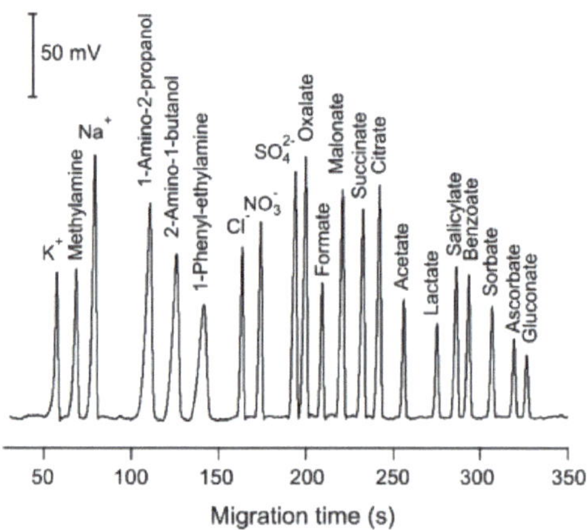

Figure 5.7: A typical electropherogram demonstrating the order of elution of cations and anions. Adapted from J. Sáiz, I. J. Koenka, T. Duc Mai, P. C. Hauser, and C. García-Ruiz, Simultaneous separation of cations and anions in capillary electrophoresis. *Trends Anal. Chem.*, 2014, 62. 162. Copyright: Elsevier (2014).

Optimizing the CE experiment

There are several components that can be varied to optimize the electropherogram obtained from CE. Hence, for any given setup certain parameters should be known:

- the total length of the capillary (L),
- the length the solutes travel from the start to the detector (l),
- the applied voltage (V).

Reduction in migration time, t_{mn}

To shorten the analysis time, a higher voltage can be used, or a shorter capillary tube can be used. However, it is important to note that the voltage cannot be arbitrarily high as it will lead to joule heating. Another possibility is to increase μ_{eof} by increasing pH or decreasing the ionic strength of the buffer,

$$t_{mn} = \frac{lL}{(\mu_{ep} + \mu_{eof})V}$$

Efficiency

In chromatography, the efficiency is given by the number of theoretical plates, N. In CE, there exist a similar parameter,

$$N = \frac{l^2}{2Dt_{mn}} = \frac{\mu_{tot}Vl}{2DL}$$

where D is the solute's diffusion coefficient. Efficiency increase s with an increase in voltage applied as the solute spends less time in the capillary there will be less time for the solute to di use. Generally, for CE, N will be very large.

Resolution between two peaks

The resolution between two peaks, R, is defined by,

$$R = \frac{\sqrt{N}}{4} \times \frac{\Delta v}{\bar{v}}$$

where Δv is the difference in velocity of two solutes and v is the average velocity of two solutes.

Substituting the equation by N gives:

$$R = 0.177 \left(\mu_{ep,1} - \mu_{ep,2}\right) \sqrt{\frac{V}{D\left(\mu_{av} + \mu_{eof}\right)}}$$

Therefore, increasing the applied voltage, V, will increase the resolution. However, it is not very effective as a 4-fold increase in applied voltage would only give a 2-fold increase in resolution. In addition, increase in N, the number of theoretical plates would result in better resolution.

Selectivity

In chromatography, selectivity, α, is defined as the ratio of the two retention factors of the solute. This is the same for CE,

$$\alpha = t_2/t_1$$

where $t2$ and $t1$ are the retention times for the two solutes such that, α is more than 1. Selectivity can be improved by adjusting the pH of the buffer solution. The purpose is to change the charge of the species being eluted.

Comparison between CE and HPLC

CE unlike High-performance liquid chromatography (HPLC) accommodates many samples and tends to have a better resolution and efficiency. A comparison between the two methods is given in Table 3.8.

CE	HPLC
Wider selection of analyte to be analyzed	Limited by the solubility of the sample
Higher efficiency, no stationary mass transfer term as there is no stationary phase	Efficiency is lowered due to the stationary mass transfer term (equilibration between the stationary and mobile phase)
Electroosmotic flow pro le in the capillary is at as a result no band broadening. Better peak resolution and sharper peaks	Rounded laminar flow pro le that is common in pressure driven systems such as HPLC. Resulting in broader peaks and lower resolution
Can be coupled to most detectors depending on application	Some detectors require the solvent to be changed and prior modification of the sample before analysis
Greater peak capacity as it uses a very large number of theoretical plates, N	The peak capacity is lowered as N is not as large
High voltages are used when carrying out the experiment	No need for high voltage

Table 5.2: Advantages and disadvantages of CE versus HPLC.

Micellar electrokinetic chromatography

CE allows the separation of charged particles, and it is mainly compared to ion chromatography. However, no separation takes place for neutral species in CE. Thus, a modified CE technique named micellar electrokinetic chromatography (MEKC) can be used to separate neutrals based on its size and its affinity to the micelle. In MEKC, surfactant species is added to the buffer solution at a concentration at which micelles will form. An example of a surfactant is sodium dodecyl sulfate (SDS, Figure 5.8). The formation of the hydrophobic interior would allow the encapsulation of the neutral species. Whilst, the negatively charge hydrophilic exterior allows the molecule to be treated as an anion. Thus, it would move towards the cathode with a velocity less than the electroosmotic flow.

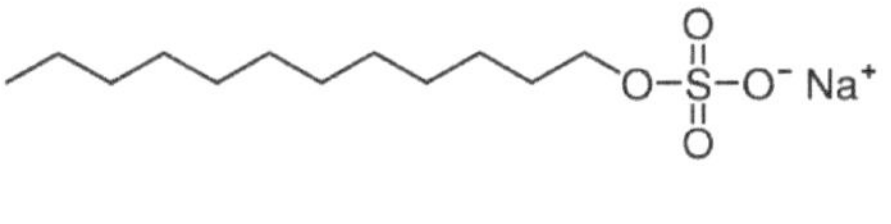

Figure 5.8: The structure of sodium dodecyl sulfate (SDS).

Neutral molecules are in dynamic equilibrium between the bulk solution and interior of the micelle. In the absence of the micelle the neutral species would reach the detector at t_0 but in the presence of the micelle, it reaches the detector at t_{mc}, where t_{mc} is greater than t_0. The longer the neutral molecule remains in the micelle, the longer it's migration time. Thus small, non-polar neutral species that favor interaction with the interior of the micelle would take a longer time to reach the detector than a large, polar species. Anionic, cationic and zwitter ionic surfactants can be added to change the partition coefficient of the neutral species. Cationic surfactants would result in positive micelles that would move in the direction of electroosmotic flow. This enables it to move faster towards the cathode. However, due to the fast migration, it is possible that insufficient time is given for the neutral species to interact with the micelle resulting in poor separation. Thus, all factors must be considered before choosing the right surfactant to be used. The mechanism of separation between MEKC and liquid chromatography is the same. Both are dependent on the partition coefficient of the species between the mobile phase and stationary phase. The main difference lies in the pseudo stationary phase in MEKC, the micelles. The micelle which can be considered the stationary phase in MEKC moves at a slower rate than the mobile ions.

The use of CE in separation of quantum dots (QD)

Quantum dots are semiconductor nanocrystals that lie in the size range of 1-10 nm, and they have different electrophoretic mobility due to their varying sizes and surface charge. CE can be used to separate and characterize such species, and a method to characterize and separate CdSe QD in the aqueous medium has been developed. The QDs were synthesized with an outer layer of trioctylphosphine (TOP, Figure 5.9) and trioctylphosphine oxide (TOPO, Figure 5.10), making the surface of the QD hydrophobic. The background electrolyte solution used was SDS, in order to make the QDs soluble in water and form a QD-TOPO/TOP- SDS complex. Different sizes of CdSe were used and the separation was with respect to the charge-to-mass ratio of the complexes. It was concluded from the study that the larger the CdSe core (i.e., the larger the charge-to-mass ratio) eluted out last. The electropherogram from the study is shown in Figure 5.11, from which it is visible that good separation had taken place by using CE. Laser-induced fluorescence detection was used, the buffer system was SDS, and the pH of the system set up was fixed at 6.5. The pH is highly important in this case as the stability of the system and the separation is dependent on it.

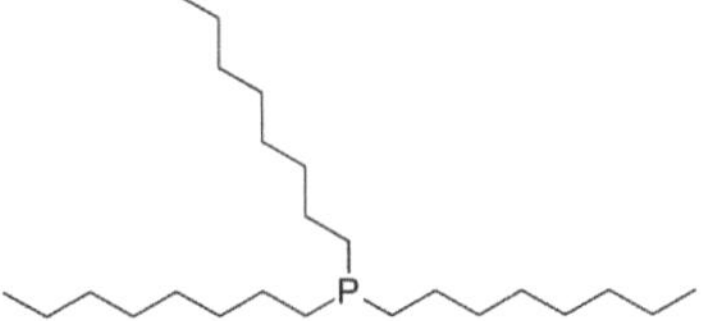

Figure 5.9: The structure of trioctylphosphine (TOP).

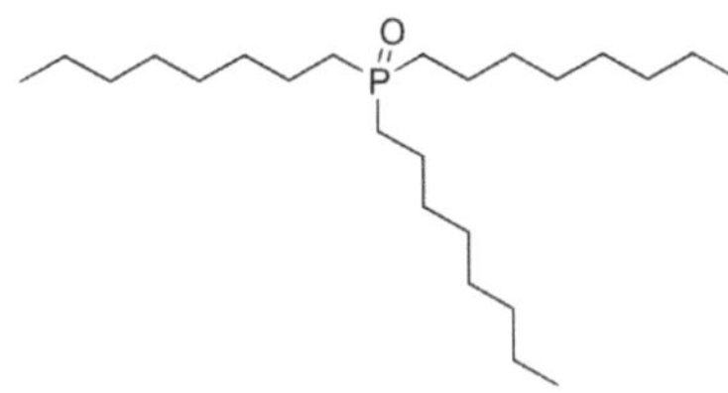

Figure 5.10: The structure of trioctylphosphine oxide (TOPO).

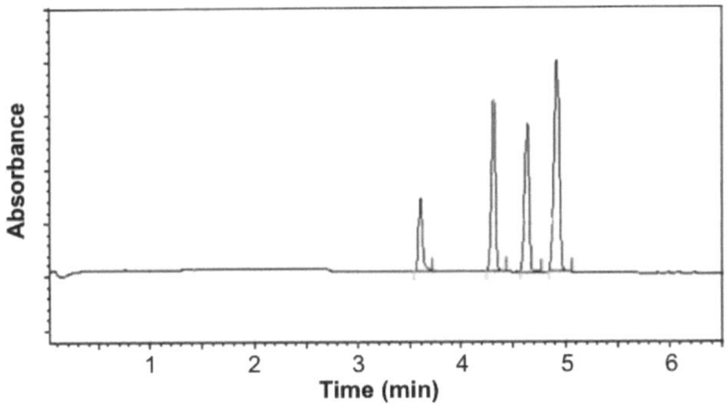

Figure 5.11: Electropherogram for a mixture of four different CdSe-TOPO/TOP-SDS complexes. Adapted from C. Carrillo-Carrión, Y. Moliner-Martínez, B. M. Simonet, and M. Valcárcel, Capillary electrophoresis method for the characterization and separation of CdSe quantum dots. *Anal. Chem.*, 2011, 83, 2807. Copyright: American Chemical Society (2011).

Bibliography

F. Rouessac and A. Rouessac, *Chemical Analysis: Modern Instrumentation Methods and Techniques*, Wiley, New York (2007).

C. Carrillo-Carrión, Y. Moliner-Martínez, B. M. Simonet, and M. Valcárcel, Capillary electrophoresis method for the characterization and separation of CdSe quantum dots. *Anal. Chem.*, 2011, **83**, 2807.

J. Sáiz, I. J. Koenka, T. Duc Mai, P. C. Hauser, C. García-Ruiz, Simultaneous separation of cations and anions in capillary electrophoresis. *Trends Anal. Chem.*, 2014, **62**. 162.

D. A. Skoog, D. M. West, F. J. Holler and S. R. Crouch, *Fundamentals of Analytical Chemistry*, Brooks Cole (2013).

D. Harvey, *Analytical Chemistry 2.0* (e-textbook), 851.

S. Hertjan, Stellan Hjertén, Uppsala University. *Analyst*, 2003, **128**, 1307.

F. Sang, X. Huang, and J. Ren, Characterization and separation of semiconductor quantum dots and their conjugates by capillary electrophoresis. *Electrophoresis*, 2014, **35**, 793.

www.ingramcontent.com/pod-product-compliance
Lightning Source LLC
Chambersburg PA
CBHW041222050726
47599CB00001B/39